CALCULATION EXAMPLES

ARITHMETIC 1

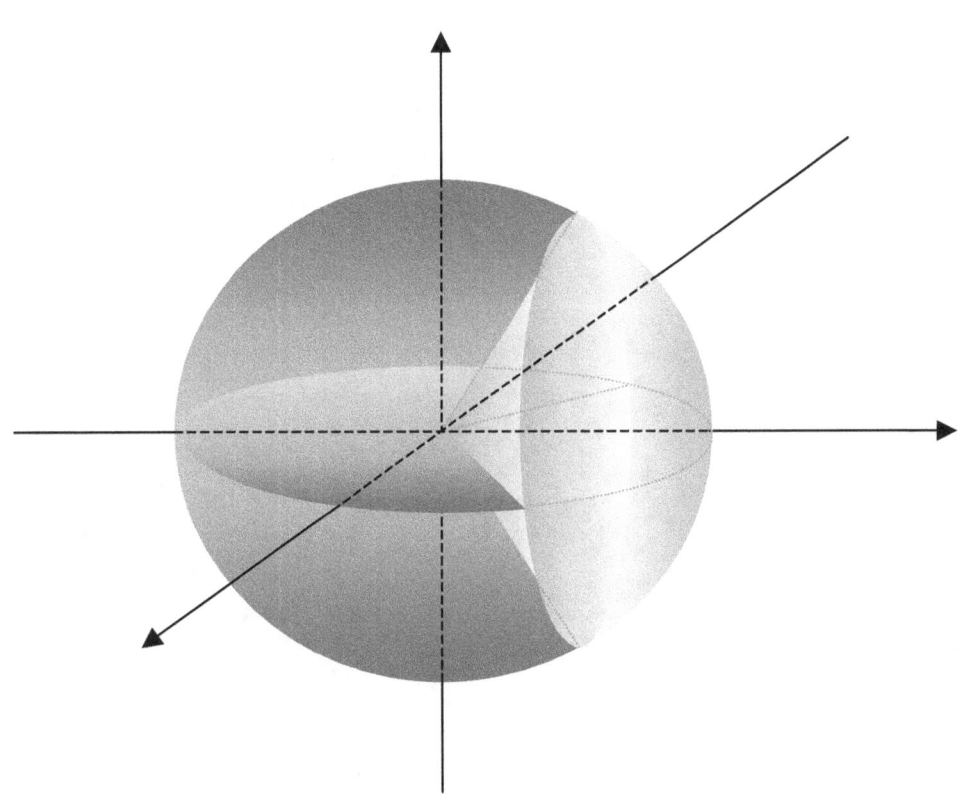

Seong R. KIM

Dear students:

Students need the best teacher, so you need examples, because examples are the best teacher. All the examples here are fully worked, and explain **how** the basic and essential tools in math are made, together with **what** they are, **how** they work, and **how** to work with them. Such tools include numbers, formulas, identities, equations, laws, etc.

Examples here begin with easy ones, of course. Covering every meter and yard properly, we can cover thousands of miles and kilometers. And it is particularly the case in math.

Of those examples therefore, some might even look too easy for you. It's not that easy though, to come up with those examples. Anyways, the bigger and the taller the tree, the deeper and the stronger the root.

Doing math, we work with ideas and run ideas, because every thing in math is an idea. A number is an idea, for instance, and the same is true for a line or circle, too. And putting ideas together, we build another, which becomes the base or an element of another, and each is connected. And that's the way your math grows. So you get to build a circuit, and sometimes, need to fill the gap or repair the circuit so that you get the sense of it.

So your calculation runs properly, and you get the problem solved.

The examples have been made and arranged so that they get tougher (or sometimes easier for some reason) as you proceed with them. In particular, similar examples with some variations are strategically repeated so that you can get the ideas or the tools tricky or complicated, and can get them mastered.

This book is however, nothing but a bunch of examples until you get it powered. How then, to get it powered, and make it run and work for you?

Just read it, and then, do each example in writing. And it is important to note that you do it in **your** writing. Just watching someone doing it, you just only feel that you can do it. If you do it, you can do it, but if you don't, we can hardly. It's a cliché, of course, but is always true that knowing is one thing and doing is another.

I've been helping students grow, take care of, and run their own math. The area covers algebra and geometry for high school or college students, and is especially for equations (for unknowns or curves), functions, and their graphs, which are the basic elements in calculus, which's been the core of my interest from my early age in high school.

Of my students, some are quite poor in math, and thus, are afraid of or hate math, some require special education because of exceptional intelligence, some are smart enough, some are naïve and diligent, some are clever but lazy, and most behave in general. All the students are badly after though, one thing in common: a strong and secure math skill. It is of course, the prime objective of my work, and I'm always happy to and eager to help them achieve it. The problem was however, that many of them wanted it to be purchased. And the question is, can we buy it?

We can buy the means, of course. And a solid math skill is feasible, too. We know however, we can't buy love, and the same is true for the math skill, too. It's not what we can buy or sell, and not what we can give or take. It is however, what we can grow, and need to grow. Your math grows as much as you grow and take care of it. So does mine.

What math then, do students most often do or use in high schools or colleges?

It is algebra and geometry. What algebra though?

Elementary algebra, of course
Doing the algebra, we work with numbers (many in kinds), constants, variables, ratios, rates, expressions, equations, inequalities, functions, identities, formulas, laws, etc., together with signs and symbols. And if we want to do algebra properly, we want to know their natures and how they mingle with each other.

So studying math ideas or tools, you want to know **what** they are, **how** they work, and **how** to work with them or **what** to do with them.   What then, about the geometry?

Basically, the geometry has much to do with shapes, positions, and angles. The shapes begin with triangles and circles, and move on to rectangles, squares, parallelograms or rhombuses, trapezoids, tetragons, other polygons, polyhedrons, etc.

Doing the geometry, too, though, we need to do the algebra stated above. So it is analytic geometry, often called coordinate geometry, too. And doing it, we can specify positions using coordinates. So in the geometry, basically, we work with graphs. Putting a math idea in a graph, we can not only effectively think about it but actually see it, too, and therefore, can efficiently work with it. What idea then, is it?

The idea begins with a point, line, parabola, circle, ellipse, and hyperbola, called a conic section or basic curve, and then, moves on to other curves, planes, surfaces, volumes, and other objects in various dimensional spaces, together with vectors.

And using an angle, we can specify an amount of turn or change in direction.

So learning, using, or applying those ideas or math tools, we get to solve problems.

And this book can help. It can help learn them, and use them so that you can navigate to find solutions to problems. And in particular, it can help come up with answers to those **what**s and **how**s stated above. So it can help you grow and run your own math, and thus, can help achieve your solid math skill.

It is however, not a magic book giving you a math skill of high caliber overnight. And it can have many mistakes, too. There is no magic, and math is full of facts and ideas. And it is after all, not me and not your teacher but you who put together some of those facts and ideas, and understand it. Putting facts and ideas together, understanding it, and taking care of what you have learned, you grow your math. And this book can help.

This is a book of examples designed to help you grow your math, and assumes that you are a real beginner. This book requires though, time and effort, the amount of which need to be substantial, too, but will be worth it. That's because you want a substantial achievement, and will get it. And probably, you will get to see this book helping you get there much faster than expected. And then, you will get to see the way math runs.

In math, everything is an idea. So is a problem. And solving it, we put it many different ways. For instance, while expanding or reducing it, or modifying or converting it, we keep searching for the solution, approaching the solution, and eventually, can get there. So don't look for the solution outside the problem. The solution is inside the problem if the problem is properly made.

If it is not, no solution is the solution. And in fact, it is often the case a problem itself is the solution. We can put a problem in many different ways, and eventually, can end up with the solution.   How come then, is the solution no other than the problem?

For instance, the solution to $3232 \div 101$ is 32.   And we can put it this way:

$$3232 \div 101 = \frac{3232}{101} = \frac{32 \times 101}{101} = \frac{32}{1} = 32 \;\Rightarrow\; 3232 \div 101 = 32.$$

And we can get this, too: $32 \Rightarrow 3232 \div 101$.   How?

$$32 = \frac{32}{1} = \frac{32 \times 101}{101} = \frac{3232}{101} = 3232/101 = 3232 \div 101. \quad \text{Too easy?}$$

For another instance, the solution to $ax^2 + bx + c = 0$ is: $x = \frac{-b \pm \sqrt{b^2 - 4ac}}{2a}$, which is called the quadratic formula.   How come then, is the solution no other than the problem?

We can put it this way:

$$x = \frac{-b \pm \sqrt{b^2 - 4ac}}{2a} \Rightarrow 2ax = -b \pm \sqrt{b^2 - 4ac} \Rightarrow 2ax + b = \pm \sqrt{b^2 - 4ac}$$

$$\Rightarrow (2ax + b)^2 = b^2 - 4ac \Rightarrow 4a^2x^2 + 4abx + b^2 = b^2 - 4ac$$

$$\Rightarrow 4a^2x^2 + 4abx = -4ac \Rightarrow ax^2 + bx = -c \Rightarrow ax^2 + bx + c = 0.$$

And we can get this, too: $ax^2 + bx + c = 0 \Rightarrow x = \frac{-b \pm \sqrt{b^2 - 4ac}}{2a}$.   How?

$$ax^2 + bx + c = a(x^2 + \tfrac{b}{a}x) + c = a(x^2 + \tfrac{b}{a}x + \tfrac{b^2}{4a^2} - \tfrac{b^2}{4a^2}) + c = a(x^2 + \tfrac{b}{a}x + \tfrac{b^2}{4a^2}) - \tfrac{b^2}{4a} + c$$

$$= a(x + \tfrac{b}{2a})^2 - \tfrac{b^2 - 4ac}{4a} = 0 \Rightarrow a(x + \tfrac{b}{2a})^2 = \tfrac{b^2 - 4ac}{4a} \Rightarrow (x + \tfrac{b}{2a})^2 = \tfrac{b^2 - 4ac}{4a^2} \Rightarrow x + \tfrac{b}{2a} = \pm\sqrt{\tfrac{b^2 - 4ac}{4a^2}}$$

$$\Rightarrow x = -\tfrac{b}{2a} \pm \tfrac{\sqrt{b^2 - 4ac}}{2a} = \tfrac{-b \pm \sqrt{b^2 - 4ac}}{2a} \Rightarrow x = \tfrac{-b \pm \sqrt{b^2 - 4ac}}{2a}.$$

And we call the set of processes above, algebra.

So if a problem is well defined, that is, if it makes sense, we should be able to get it solved the way below:

A problem $\Rightarrow$ **...** $\Rightarrow$ **...** $\Rightarrow$ **the solution**, and thus: **the problem** $\Rightarrow$ **the solution**.

So solving a problem, we put it many different ways so that we can get to the solution.

And that's the way, math runs.

May your math run very well.

Seong R. Kim

B.S. Math. Michigan Tech. Univ.   M.S. Math. Rensselaer Polytechnic Institute

Notes:

This book is one of three books about some basics in numbers, and covers arithmetic operations. And the operations are on numbers called real numbers. We can classify real numbers in two ways. In one, a number is an integer or a non-integer, which is not an integer. And in the other, a number is rational or irrational, which is not rational.

So if a number is not irrational, it is rational. Among rational numbers, we have integers and non-integers. And if a number is not an integer, it is a non-integer. Among non-integers, we have rational numbers and irrational numbers. It's confusing, isn't it?

So usually, we classify real numbers into three groups: integers, rational numbers, and irrational numbers.

And this book covers integers, and helps understand what integers are about and how they mingle with arithmetic operations so that you can not only use integers properly but can develop your own idea to make use of them, too, solving problems, of course.

It doesn't just cover though, how to do arithmetic with integers, that is, how to do additions, subtractions, multiplications, and divisions. But it helps also understand the nature of integers and the nature of arithmetic operations as additions or divisions. And in particular, it covers the nature and the use of numbers called inverses. What inverse?

It can be called the negative or the reciprocal depending on the arithmetic operation. So this book covers those math ideas (concepts), and helps you understand them, and use them efficiently as well as properly. It can help speed up your calculation, too.

What then, about the other numbers?

They are covered in **Calculation Examples Arithmetic 3**, which covers thus, numbers rational as 0.5, -3/2, or 5/3, and numbers irrational as a square root of 2.

What then, about **Calculation Examples Arithmetic 2**?

The book is designed for those students who want to study word problems, where students need to manage integers and set up expressions. And you will get to see some nature of integers and some tools useful or handy using or working with integers. The tools are math ideas, called theorems or formulas, which you can use doing problems with multiples or divisors, etc.

And thus, all the three books will help you grab math ideas often used in real life as well as in math courses. The ideas are about real numbers, their arithmetic operations, and their nature, so you will get to see what those numbers are about and how they mingle with arithmetic operations, and will get to develop your own idea to make use of those, solving problems, of course.

In short, the books help you strengthen fundamentals in math to increase your skill of algebra, that is, calculation techniques, providing you with examples, showing all the steps and the ideas behind, and explaining what the math ideas are about.

Contents

Examples 1

Answer the questions below:

(The first some questions may seem too easy.)

0.   How many 1s do we need to make 5?

1.   How many 1s do we need to make 12?

2.   How many 1s do we need to make 45?

3.   How many 1s do we need to make 283?

4.   How many 5s do we need to make 5?

5.   How many 5s do we need to make 10?

6.   How many 5s do we need to make 0?

7.   How many 5s and 1s do we need to make 7?

8.   How many 3s and 2s do we need to make 7?

9. How many 3s, 2s, and 1s do we need to make 7?

A. How many 10s do we need to make 80?

B. How many 10s do we need to make 100?

C. How many 10s do we need to make 190?

D. How many 10s do we need to make 350?

E. How many 10s do we need to make 990?

F. How many 10s do we need to make 1000?

G. How many 10s do we need to make 1040?

H. How many 10s do we need to make 1520?

I. How many 10s do we need to make 3500?

J. How many 10s do we need to make 6390?

K. How many 100s do we need to make 900?

L. How many 100s do we need to make 1000?

M. How many 100s do we need to make 9000?

N. How many 100s do we need to make 3800?

O. So what pattern can you see in the sequence of all the answers to the questions asking the numbers of 10s or 100s above?

Suggestions or Solutions
To the Examples 1

0.   How many 1s do we need to make 5?

5 has 5 of 1s, so we get: 5 x 1 = 5.

1.   How many 1s do we need to make 12?

12 has 12 of 1s, and we get: 12 x 1 = 12

2.   How many 1s do we need to make 45?

45 has 45 of 1s, so we get: 45 x 1 = 45

3.   How many 1s do we need to make 283?

283 has 283 of 1s, and we get: 283 x 1 = 283

4.   How many 5s do we need to make 5?

5 has 1 of 5s, so we get: 1 x 5 = 5

5.   How many 5s do we need to make 10?

10 has 2 of 5s, and we get: 2 x 5 = 10

6.   How many 5s do we need to make 0?

0 has none of 5s, so we get: 0 x 5 = 0.

7.   How many 5s and 1s do we need to make 7?

7 has 1 of 5s and 2 of 1s, so we get: 1 x 5 + 2 x 1 = 7.   We can get it this way, too, though: 0 x 5 + 7 x 1 = 7.

8.   How many 3s and 2s do we need to make 7?

7 has 1 of 3s and 2 of 2s, and we get: 1 x 3 + 2 x 2 = 7, and we can be put it this way, too, of course: 0 x 3 + 3 x 2 + 1 x 1 = 2 x 3 + 0 x 2 + 1 x 1 = ...

9.   How many 3s, 2s and 1s do we need to make 7?

7 has 1 of 3s, 1 of 2s, and 2 of 1s, so we get: 1 x 3 + 1 x 2 + 2 x 1 = 7, and it can be put this way, too, of course: 1 x 3 + 2 x 2 + 0 x 1 = 0 x 3 + 3 x 2 + 1 x 1 = ...

A.   How many 10s do we need to make 80?

80 has 8 of 10s, so we get: 8 x 10 = 80.

B.   How many 10s do we need to make 100?

100 has 10 of 10s, so we get: 10 x 10 = 100.

C.   How many 10s do we need to make 190?

190 has 19 of 10s, and we get: 19 x 10 = 190.

D. How many 10s do we need to make 350?

350 has 35 of 10s, so we get: 35 x 10 = 350.

E. How many 10s do we need to make 990?

990 has 99 of 10s, so we get: 99 x 10 = 990.

F. How many 10s do we need to make 1000?

1000 has 100 of 10s, so we get: 100 x 10 = 1000

G. How many 10s do we need to make 1040?

1040 has 104 of 10s, and we get: 104 x 10 = 1040.

H. How many 10s do we need to make 1520?

1520 has 152 of 10s, so we get: 152 x 10 = 1520.

I. How many 10s do we need to make 3500?

3500 has 350 of 10s, and we get: 350 x 10 = 3500.

J. How many 10s do we need to make 6390?

6390 has 639 of 10s, and we get: 639 x 10 = 6390.

K. How many 100s do we need to make 900?

900 has 9 of 100s, so we get: 9 x 100 = 900.

6

L.   How many 100s do we need to make 1000?

1000 has 10 of 100s, and we get: 10 x 100 = 1000.

M.   How many 100s do we need to make 9000?

9000 has 90 of 100s, so we get: 90 x 100 = 9000.

N.   How many 100s do we need to make 3800?

3800 has 38 of 100s, and we get: 38 x 100 = 3800

O.   What pattern can you see in the sequence of all the correct answers to the questions asking the numbers of 10s or 100s above?

Every time we multiply a particular number by 10, we get the product the way below:

All the individual digits in the particular number get shifted to one-step higher in digit. Therefore, if the particular number is an integer, the 1's digit in the product is 0. In other words, we put (append) a zero next to the 1's digit of the particular number when we get the product of the particular number and 10.

So for instance, taking the product of 24 and 100, we put (append) two zeros next to the 1's digit of the integer 24. Then, we get: 2400. What if we multiply 12 by 300?

We know: 300 = 3 x 100.
So the product is three times the product of 12 and 100.
And thus, taking the product of 12 and 3, and putting two zeros next to the 1's digit in the product, we get the product of 12 and 300, which is 3600.

Now, what if we divide an integer by 10, 100, or 1000?

Examples 2

Answering the questions below, use a number: 7406.

0. How many 10s are there in the number?

1. How many 1s are there in the number?

2. How many 100s are there in the number?

3. How many 1000s are there in the number?

4. How many 1s are there in the 10's digit?

5. How many 1s are there in the 1's digit?

6. How many 1s are there in the 1000's digit?

7. How many 1s are there in the 100's digit?

8. How many 3s are there in 1's digit?

9. How many 2s are there in 100's digit?

A. How many 7s are there in the 1000's digit?

B. How many 4's are there in the 10's digit?

Suggestions or Solutions
To the Examples 2

We are given: 7406.

0. How many 10s are there in the number 7406?

7406 = 740 x 10 + 6, so it has 740 of 10s.

1. How many 1s are there in the number?

7406 = 7406 x 1, so it has 7406 of 1s.

2. How many 100s are there in the number?

7406 = 74 x 100 + 6, and thus, it has 74 of 100s.

3. How many 1000s are there in the number?

7406 = 7 x 1000 + 406, so it has 7 of 1000s.

4. How many 1s are there in the 10's digit in the number?

10's digit has 0. So the 10's digit has none of 1s.

5.   How many 1s are there in the 1's digit in the number, 7406, of course?

1's digit has 6. So it has 6 of 1s.

6.   How many 1s are there in the 1000's digit in the number?

1000's digit has 7. So it has 7 of 1s.

7.   How many 1s are there in the 100's digit?

100's digit has 4. So it has 4 of 1s.

8.   How many 3s are there in 1's digit?

1's digit has 6. And 6 = 2 x 3. So it has 2 of 3s.

9.   How many 2s are there in 100's digit?

100's digit has 4. And 4 = 2 x 2. So it has 2 of 2s.

A.   How many 7s are there in the 1000's digit?

1000's digit has 7. And 7 = 1 x 7. So it has 1 of 7s.

B.   How many 4's are there in the 10's digit?

10's digit has 0. And 0 = 0 x 4. So it has none of 4s.

Examples 3

0. What is a number made of 2 of 1s, 5 of 10s, and 7 of 100s?

1. What number has none of 1s, 4 of 10s, and 9 of 100s?

2. If a number has 9 of 10s, 3 of 1s, and 12 of 100s, what is the number?

3. What is a number that has 14 of 1s, 9 of 10s, 19 of 100s, and 19 of 1000s?

4. What is a number has 12 of 10s, 215 of 1s, none of 100s, and 121 of 1000s?

5. What is a number has 2 of 0.1s, 5 of 0.01s, and 7 of 0.001s?

6. What number has none of 0.1s, 4 of 0.01s, 7 of 0.001s, and 9 of 0.0001s?

7. If a number has 9 of 0.1s, 13 of 0.01s, and 24 of 0.001s, what is the number?

8. What number has 14 of 0.1s, 19 of 0.01s, 91 of 0.001s, and 19 of 0.0001s?.

9. What number has 27 of 0.1s, 215 of 0.01s, none of 0.001s, and 151 of 0.0001s?

Suggestions or Solutions
To the Examples 3

0. What is a number made of 2 of 1s, 5 of 10s, and 7 of 100s?

2 x 1 + 5 x 10 + 7 x 100 = 752

1. What number has none of 1s, 4 of 10s, and 9 of 100s?

0 x 1 + 4 x 10 + 9 x 100 = 940

2. If a number has 9 of 10s, 3 of 1s, and 12 of 100s, what is the number?

9 x 10 + 3 x 1 + 12 x 100 = 3 + 90 + 1200 = 3 + 90 + 200 + 1000 = 1293

3. What is a number that has 14 of 1s, 9 of 10s, 19 of 100s, and 19 of 1000s?

14 x 1 + 9 x 10 + 19 x 100 + 19 x 1000
= 14 + 90 + 1900 + 19000
= 4 + 10 + 90 + 900 + 1000 + 9000 + 10000
= 4 + 100 + 900 + 10000 + 10000
= 4 + 1000 + 20000
= 21004

4. What is a number has 12 of 10s, 215 of 1s, none of 100s, and 121 of 1000s?

12 x 10 + 215 x 1 + 0 x 100 + 121 x 1000
= 120 + 215 + 121000
= 100 + 20 + 200 + 10 + 5 + 100000 + 20000 + 1000
= 100000 + 20000 + 1000 + 300 + 30 + 5
= 121335

5. What is a number has 2 of 0.1s, 5 of 0.01s, and 7 of 0.001s?

2 x 0.1 + 5 x 0.01 + 7 x 0.001
= 0.2 + 0.05 + 0.007 = 0.257

6. What number has none of 0.1s, 4 of 0.01s, 7 of 0.001s, and 9 of 0.0001s?

0 x 0.1+ 4 x 0.01 + 7 x 0.001 + 9 x 0.0001
= 0 + 0.04 + 0.007 + 0.0009 = 0.0479

7. If a number has 9 of 0.1s, 13 of 0.01s, and 24 of 0.001s, what is the number?

9 x 0.1 + 13 x 0.01 + 24 x 0.001
= 0.9 + 0.13 + 0.024
= 0.9 + 0.1 + 0.03 + 0.02 + 0.004
= 1 + 0.05 + 0.004
= 1.054

14

8. What number has 14 of 0.1s, 19 of 0.01s, 91 of 0.001s, and 19 of 0.0001s?.

14 x 0.1 + 19 x 0.01 + 91 x 0.001 + 19 x 0.0001

= 1.4 + 0.19 + 0.091 + 0.0019

= 1 + 0.4 + 0.1 + 0.09 + 0.09 + 0.001 + 0.001 + 0.0009

= 1 + 0.5 + 0.18 + 0.002 + 0.0009

= 1 + 0.5 + 0.1 + 0.08 + 0.002 + 0.0009

= 1 + 0.6 + 0.08 + 0.002 + 0.0009

= 1.6829

9. What number has 27 of 0.1s, 215 of 0.01s, none of 0.001s, and 151 of 0.0001s?

27 x 0.1 + 215 x 0.01 + 0 x 0.001 + 151 x 0.0001

= 2.7 + 2.15 + 0 + 0.0151

= 2 + 0.7 + 2 + 0.1 + 0.05 + 0.01 + 0.005 + 0.0001

= 4 + 0.8 + 0.06 + 0.005 + 0.0001

= 4.8651

Examples 4

Find the sum in each of the cases below:

0. 3 of 1000s, 2 of 100s, 4 of 10s, and 5 of 1s

1. 7 of 1000s, 12 of 100s, 16 of 10s, and 15 of 1s

2. 18 of 1000s, 4 of 100s, and 128 of 1s

3. 12 of 1000s, 724 of 10s, and 309 of 1s

4. 27 of 1000s, 103 of 100s, 802 of 10s, and 109 of 1s

5. 308 of 1000s, 207 of 100s, 1003 of 10s, and 104 of 1s

6. 2 of 1s, 52 of 0.01s, and 75 of 0.001s

7. 7 of 10s, 17 of 1s, 45 of 0.1s, 72 of 0.01s, and 902 of 0.001s

8. 12 of 10s, 127 of 1s, 95 of 0.1s, 143 of 0.01s, and 294 of 0.001s

9. 14 of 10s, 195 of 0.1s, 921 of 0.001s, and 109 of 0.0001s

A. 207 of 1s, 2115 of 0.01s, 1000 of 0.001s, and 1501 of 0.0001s

Suggestions or Solutions
To the Examples 4

0. **3 of 1000s, 2 of 100s, 4 of 10s, and 5 of 1s**

3 x 1000 + 2 x 100 + 4 x 10 + 5 x 1 = 3245

1. **7 of 1000s, 12 of 100s, 16 of 10s, and 15 of 1s**

7 x 1000 + 12 x 100 + 16 x 10 + 15 x 1
= 7000 + 1200 + 160 + 15
= 7000 + 1000 + 200 + 100 + 60 + 10 + 5
= 8000 + 300 + 70 + 5 = 8375

2. **18 of 1000s, 4 of 100s, and 128 of 1s**

18 x 1000 + 4 x 100 + 128 x 1
= 18000 + 400 + 128
= 18000 + 400 + 100 + 28
= 18000 + 500 + 28 = 18528

3. **12 of 1000s, 724 of 10s, and 309 of 1s**

12 x 1000 + 724 x 10 + 309 x 1
= 12000 + 7240 + 309
= 10000 + 2000 + 7000 + 200 + 40 + 300 + 9
= 10000 + 9000 + 500 + 40 + 9
= 19549

4. 27 of 1000s, 103 of 100s, 802 of 10s, and 109 of 1s

27 x 1000 + 103 x 100 + 802 x 10 + 109 x 1
= 27000 + 10300 + 8020 + 109
= 20000 + 7000 + 10000 + 300 + 8000 + 20 + 100 + 9
= 30000 + 15000 + 400 + 29
= 45000 + 429
= 45429

5. 308 of 1000s, 207 of 100s, 1003 of 10s, and 104 of 1s

308 x 1000 + 207 x 100 + 1003 x 10 + 104 x 1
= 308000 + 20700 + 10030 + 104
= 300000 + 8000 + 20000 + 700 + 10000 + 30 + 100 + 4
= 300000 + 30000 + 8000 + 800 + 34
= 338834

6. 2 of 1s, 52 of 0.01s, and 75 of 0.001s

2 x 1 + 52 x 0.01 + 75 x 0.001
= 2 + 0.52 + 0.075
= 2 + 0.5 + 0.02 + 0.07 + 0.005
= 2 + 0.5 + 0.09 + 0.005
= 2.595

7. 7 of 10s, 17 of 1s, 45 of 0.1s, 72 of 0.01s, and 902 of 0.001s

7 x 10 + 17 x 1 + 45 x 0.1 + 72 x 0.01 + 902 x 0.001
= 70 + 17 + 4.5 + 0.72 + 0.902
= 70 + 10 + 7 + 4 + **0.5** + **0.7** + 0.02 + **0.9** + 0.002
= 80 + 11 + **21 x 0.1** + 0.02 + 0.002
= 80 + 10 + 1 + 2.1 + 0.02 + 0.002
= 90 + 1 + 2 + 0.1 + 0.02 + 0.002
= 90 + 3 + 0.1 + 0.02 + 0.002
= 93.122

8. 12 of 10s, 127 of 1s, 95 of 0.1s, 143 of 0.01s, and 294 of 0.001s

12 x 10 + 127 x 1 + 95 x 0.1 + 143 x 0.01 + 294 x 0.001
= 120 + 127 + 9.5 + 1.43 + 0.294
= 247 + 9 + 0.5 + 1 + 0.4 + 0.03 + 0.2 + 0.09 + 0.004
= 247 + 10 + 11 x 0.1 + 12 x 0.01 + 0.004
= 257 + 1.1 + 0.12 + 0.004
= 257 + 1 + 0.1 + 0.124
= 258 + 0.224
= 258.224

9. 14 of 10s, 195 of 0.1s, 921 of 0.001s, and 109 of 0.0001s

14 x 10 + 195 x 0.1 + 921 x 0.001 + 109 x 0.0001
= 140 + 19.5 + 0.921 + 0.0109
= 140 + 10 + 9 + 0.5 + 0.9 + 0.02 + 0.001 + 0.01 + 0.0009
= 159 + 14 x 0.1 + 0.03 + 0.0019
= 159 + 1.4 + 0.0319
= 159 + 1 + 0.4 + 0.0319
= 160.4319

A. 207 of 1s, 2115 of 0.01s, 1000 of 0.001s, and 1501 of 0.0001s

207 x 1 + 2115 x 0.01 + 1000 x 0.001 + 1501 x 0.0001
= 207 + 21.15 + 1 + 0.1501
= 200 + 7 + 20 + 1 + 0.15 + 1 + 0.15 + 0.0001
= 229 + 2 x 0.15 + 0.0001
= 229 + 0.3 + 0.0001
= 229.3001

1. The Basics 1

To begin with, doing math, we use numbers, of course, and they are most often used among all the things in math. And of all the numbers we use, the most basic and the most often used are integers. Using integers in fact, we make numbers of other kinds, too, together with signs or symbols.   What is an integer though?

The word 'integer' is probably a composite word made of 'integral' and 'number'. That is to say that **integer =** <u>**integ**</u>ral + numb<u>**er**</u>.

So an integer can be called an integral number, and can be a whole number as 1, 2, 3, etc. called a natural number, too, and thus, can indicate an amount as one or three, and does not indicate any amount with a fraction as a half, three halves, two thirds, or seven fifths.

And in integers, we have positive ones and negative ones, together with 0.

So we have: 0, positive integers, and negative integers.

We can use a whole number as a positive integer. So for instance, a positive integer can be 2 or 9. Though it is positive, we don't usually use the plus sign, so we take 2 as +2.

What then, do we mean by a negative integer?

A negative integer is a product of –1 and an integer positive, and can be taken as the number opposite of the integer positive, because the two have opposite signs.

For instance, a negative integer can be: –2, and we can get it doing this: (–1) x 2 or this, of course: 2 x (–1), and we can get –9, doing this: (–1) x 9.

So we have: –2 = (–1) x 2, and –9 = (–1) x 9.
And the sign of 2, that is, the sign of +2 is the opposite of the sign of –2.

By the way, doing a multiplication, we don't usually use the operator x, and instead, we just use a dot, or nothing if no ambiguity is expected.

So we often just put the expression above this way: $-2 = (-1)\cdot 2$, or $-9 = (-1)\cdot 9$.

And -2 can be taken as the number opposite of 2, and -9 can be taken as the number opposite of 9.

Formally though, a negative integer is called the additive inverse of a positive integer. And vice versa. So a positive integer is the additive inverse of a negative integer.

What do we mean by the additive inverse though?

The sum of an integer and its additive inverse is 0.
So if two integers add up to 0, what is one of the two to the other?

It's the additive inverse of the other. So the two are additive inverse of each other.

So for instance, 2 and -2 are additive inverse of each other, and we get: $2 + (-2) = 0$, which is thus, the same as: $2 - 2$, so we have: $2 + (-2) = 2 - 2$.

What does then, the additive inverse have to do with a negative integer?

The additive inverse of an integer is the negative of the integer.
So is the negative of an integer, negative?

Not necessarily

The negative of an integer can be positive, 0, or negative.

If an integer is positive, the negative of the integer is negative. If an integer is negative, its negative is positive. And if it is 0, its negative is 0, too.

So for instance, the negative of 5 is: -5, the negative of -5 is: $-(-5) = 5$.
And the negative of 0 is 0, so the negative of an integer is not always negative.

How then, can we make a positive integer negative?

Multiplying or dividing an integer by -1, we get <u>the negative of</u> the integer.

And if the integer is positive, the negative of the integer is negative.

So for instance, the negative of 23 is: -1 x 23 = -23, or 23/(-1) = -23, which is a negative integer, and is the opposite of 23, that is, the additive inverse of 23.

Thus, we get: -23 + 23 = 0, and 23 + (-23) = 0, that is, 23 + (-23) = 23 – 23 = 0.

How then, can we make a negative integer positive?

We know multiplying or dividing an integer by –1, we get the negative of the integer.

And if the integer is negative, the negative of the integer is positive.

So for instance, the negative of -34 is -1 x (-34) = -(-34) = 34, or (-34)/(-1) = 34, which is a positive integer, and is the opposite of -34, that is, the additive inverse of -34.

Thus, we get: 34 + (-34) = 0, that is, 34 + (-34) = 34 – 34 = 0, and (-34) + 34 = 0, too.

So multiplying or dividing an integer by –1, we <u>change the sign</u> of the integer.

Thus, multiplying or dividing a *positive* integer by –1, we make the integer *negative*.
And multiplying or dividing a *negative* integer by –1, we make the integer *positive*.

So how do we make a negative integer?

We can make it multiplying a positive integer by –1, and of course, dividing the positive integer by –1, too.
In fact, every nonzero integer has its negative. And taking the sum of an integer and its negative, we get 0, so both are equal in magnitude and are opposite in sign.

And we call the magnitude, the absolute value, too, which is positive or 0.

So the magnitude of an integer is the absolute value of the integer, which is greater than or equal to 0, that is, $\geq \mathbf{0}$. And indicating the absolute value of an integer, we use as a sign a pair of vertical bars, between which the integer is placed.

For instance, the absolute value of -7 is indicated by $|-7|$, which is thus, the magnitude of -7, and is 7. And the magnitude or the absolute value of 7 is 7, too, of course.

That is to say that $|-7| = |7| = 7$.   And of course, we have: $|0| = 0$.

And the same is true for any other kind in numbers, too.   So what do we mean by a negative number?

A negative number is a product of -1 and a number positive, and can be taken as the number opposite of the number positive.

So for instance, a negative number can be: -2, which is -1 times 2, that is, $-2 = (-1) \cdot 2$, and more examples can be: $-0.5 = (-1) \cdot 0.5$, $-\sqrt{3} = (-1)\sqrt{3}$, and $-\frac{2}{3} = (-1) \cdot \frac{2}{3} = \frac{-2}{3} = \frac{2}{-3}$.

And thus, -2 can be taken as the number opposite of 2, and 0.5 can be taken as the number opposite of -0.5.

So in sum, integers are whole numbers as 2, 0, and -5, and we have three kinds, one is positive, another is 0, and the other is negative.

And we can change the sign of a number multiplying (dividing) by -1.

And adding together two numbers opposite of each other, we get 0.
The two are equal in magnitude, but opposite in sign.
One of the two is positive, and the other is negative. So 0 can be said to be neutral.

Formally though, the two are said to be additive inverse of each other.

And in fact, subtracting a number, we add its negative, that is, the additive inverse of it. So for instance, subtracting 2, we can add -2, and subtracting -3, we add $-(-3)$, that is, 3.

So doing a subtraction, we actually do an addition adding the additive inverse, that is, the negative of the number subtracted.

And we know 0 is nether negative nor positive, that is, neutral.

So adding together numbers negative and positive, we just keep neutralizing using the idea of the additive inverse. For instance, we can do a subtraction the way below:

$5 - 2 = 3 + 2 - 2 = 0$, and the idea behind is: $5 - 2 = 5 + (-2) = 3 + 2 + (-2) = 3 + 0 = 3$.

And we can put it this way, too: $5 - 2 = -2 + 5 = -2 + 2 + 3 = 0 + 3 = 3$.

So in the calculation above, -2 gets neutralized with 2, and vice versa.

And the additive inverse of -9 is 9, that is, the negative of -9 is 9.
So for instance, we can get: $5 - (-9) = 5 + 9 = 14$.

And for another instance, we know: $\textbf{7a = 5a + 2a}$, so we can get:

$\textbf{7a} - \textbf{2a} = \textbf{5a + 2a} - \textbf{2a}$, and the idea behind is as follows:

$\textbf{7a} - \textbf{2a} = \textbf{7a + (-2a) = 5a + 2a + (-2a) = 5a + 0 = 5a}$.

And we can put it this way, too: $\textbf{-2a + 7a = -2a + 2a + 5a = 0 + 5a = 5a}$.

So in the calculation above, $\textbf{-2a}$ gets neutralized with $\textbf{2a}$, and vice versa.

And of course, the additive inversed of $\textbf{-a}$ is $\textbf{a}$, that is, the negative of $\textbf{-a}$ is $\textbf{a}$.
So for instance, we can get: $\textbf{4a} - \textbf{(-a) = 4a + a = 5a}$.

It's a good idea therefore, to get used to the idea of the additive inverse, that is, the negative of a number or an expression.

So in the next section, we will go over the basics above, the idea of the additive inverse, and then, add some more basics, after doing some examples in the next several pages.

What then, about multiplications and divisions?

The idea applies to subtractions applies to divisions, too.

So doing a division, we can do a multiplication.
When we do a division in fact, what we actually do is a multiplication.
Without doing a multiplication, we cannot do a division.

Doing a division, we can multiply by an inverse, which is thus, called the multiplicative inverse. Usually though, we call it the reciprocal.

So dividing by a number, we multiply by the multiplicative inverse called the reciprocal.

And we will get to cover the idea of the reciprocal in the section, The Basics 3.

Examples 5

Find the sums below:

0. 10 + (-9)

1. -7 + 14

2. 5 + (-8)

3. -4 + 10

4. -8 + 3

5. 18 + (-12)

6. -15 + 19

7. 19 + (-7)

8. 47 + (-39)

9. -396 + 207

A. -967 + 189

B. 628 + (-799)

C. -1937 + 859

D. 9572 + (-40001)

Suggestions or Solutions
To the Examples 5

Note that you don't have to do calculations the way below. There can be many ways to calculate.

Find the sums below:

0. 10 + (-9)

$10 = 9 + 1$, so we get: $10 + (-9) = 1 + 9 + (-9) = 1 + 0 = 1$.
And of course, we can do it this way, too: $10 - 9 = 1$.

1. -7 + 14

$14 = 7 + 7$, so we get: $(-7) + 14 = (-7) + 7 + 7 = 0 + 7 = 7$

2. 5 + (-8)

$-8 = -3 + (-5) = -3 - 5$.
So we get: $5 + (-8) = 5 + \{-3 + (-5)\} = 5 + (-3 - 5) = 5 - 3 - 5 = 0 - 3 = -3$

3. -4 + 10

$10 = 4 + 6$, so we get: $-4 + 10 = -4 + 4 + 6 = 0 + 6 = 6$

4. -8 + 3

$-8 = -5 + (-3) = -5 - 3$, so we get: $-8 + 3 = -5 - 3 + 3 = -5$

5. 18 + (-12)

$18 = 16 + 2$, so $18 + (-12) = 6 + 12 + (-12) = 6 + 0 = 6$

6.   -15 + 19

19 = 15 + 4, so -15 + 19 = -15 + 15 + 4 = 0 + 4 = 4

7.   19 + (-7)

19 = 12 + 7, so 19 + (-7) = 12 + 7 + (-7) = 12.

8.   47 + (-39)

47 = 39 + 8, so 47 + (-39) = 39 + (-39) + 8 = 8

9.   -396 + 207

-396 = -296 – 100 = -206 – 190

So -396 + 207 = -206 – 190 + 207 = -206 – 1 + 1 – 190 + 207 = 1 – 190 = -189

And we can get the same this way, too:

-396 + 207 = -200 – 196 + 200 + 7 = -196 + 7 = -190 – 6 + 7 = -189 – 1 – 6 + 7

= -189 – 7 + 7 = -189

A.   -967 + 189

-967 = -900 – 60 – 7, and 189 = 100 + 80 + 9

-967 + 189 = -800 + 20 + 2 = -700 – 100 + 20 + 2 = -700 – 80 + 2

= -700 – 70 – 10 + 2 = -700 – 70 – 8 = -778

B.   628 + (-799)

-799 = -800 + 1
628 + (-799) = 628 – 800 + 1 = 600 + 28 – 600 – 200 + 1 = -200 + 29
= -100 – 100 + 20 + 9 = -100 – 80 + 9 = -100 – 70 – 10 + 9 = -171

And we can put it this way, too:

628 = 630 – 2
628 + (-799) = 630 – 2 – 799 = 630 – 1 – 1 – 799 = 630 – 1 – 800
= 600 + 30 – 1 – 600 – 200 = 30 – 1 – 200 = -170 – 1 = -171

C.   -1937 + 859

= -1900 – 40 + 3 + 800 + 60 – 1
= -1100 + 20 + 2
= -1000 – 100 + 20 + 2
= -1000 – 80 + 2
= -1000 – 78
= -1078

D.   9572 + (-40001)

9572 = 9580 – 8 = 9600 – 20 – 8 = 10000 – 428 = 10000 – 430 + 2
= 10000 – 500 + 72 = 10000 – 500 + 80 – 8

-40001 = -40000 – 1

9572 + (-40001)
= 10000 – 500 + 80 – 8 – 40000 – 1
= -30000 – 500 + 80 – 9
= -30000 – 420 – 9 = -30429

Note again, you don't have to do calculations the way above. There can be many ways to get the same.

2. The Basics 2

To begin with, if we get 0 adding together two numbers, the two are opposite of each other. And the two are equal in magnitude but opposite in sign. One is positive, and the other is negative.

Next, we can change the sign of a number multiplying (or dividing) it by −1.
So multiplying a positive number by −1, we get a number negative.
And multiplying a negative number by −1, we get a number positive.

And also, a negative number can be called the <u>additive inverse of a positive number</u>.

And vice versa. So a positive number is the additive inverse of a negative number.

What do we mean by the additive inverse though?

A number and its additive inverse add up to 0.
That is, adding together a number and its additive inverse, we get 0.
So what can we call the additive inverse?

The <u>additive inverse of a number</u> can be called the <u>negative of the number</u>.

The negative of an integer can be positive, 0, or negative.

If a number is positive, the negative of the number is negative.
If a number is negative, its negative is positive. And if it is 0, its negative is 0, too.

So for instance, the negative of 0.5 is: –0.5, the negative of –0.5 is: –1(–0.5) = 0.5.
And the negative of 0 is 0, so the negative of a number is not always negative.
How then, can we make a positive number negative?

Multiplying or dividing a number by –1, we get the negative of the number.
And if the number is positive, the negative of the number is negative.
So multiplying a positive number by –1, we make it negative.

So for instance, the negative of 2.3 is -1 x 2.3 = -2.3, which is a negative number, and is
the opposite of 2.3, that is, the additive inverse of 2.3.

Thus, we get: -2.3 + 2.3 = 0, and 2.3 + (-2.3) = 0, that is, 2.3 + (-2.3) = 2.3 – 2.3 = 0.
How then, can we make a negative number positive?

Multiplying or dividing a number by –1, we get the negative of the number.
And if the number is negative, the negative of the number is positive.
So multiplying a negative number by –1, we make it positive.

So for instance, the negative of -3.4 is: -1 x (-3.4) = -(-3.4) = 3.4, or (-3.4)/(-1) = 3.4
which is a positive number, and is the opposite of -3.4, i.e., the additive inverse of -3.4.

Thus, we get: 3.4 + (-3.4) = 0, that is, 3.4 + (-3.4) = 3.4 – 3.4 = 0, and (-3.4) + 3.4 = 0.
So multiplying or dividing a number by -1, we change the sign of the number.

Thus, multiplying or dividing a *positive* number by -1, we make the number *negative*.
And multiplying or dividing a *negative* number by -1, we make the number *positive*.

So how do we make a negative number?

We can make it multiplying a positive number by -1, and of course, dividing the positive
number by -1, too.

And multiplying a number by –1, we get the negative of the number, and add a negative sign to the number. For instance, -1 x 2 = -2, that is, -2 = -1 x 2, and -1 x (-3) = -(-3) = 3, because 3 is the negative of -3.

And in fact, every nonzero number has its negative, and taking the sum of a number and its negative, we get 0. And both are equal in magnitude and are opposite in sign.

And we call the magnitude <u>the absolute value</u>, too, which is positive or 0.

So the magnitude of a number is the absolute value of the number, which is greater than or equal to 0, that is, $\geq \mathbf{0}$. And indicating the absolute value of a number, we use as a sign <u>a pair of vertical bars</u>, between which the number is placed.

For instance, the absolute value of -0.7 is indicated by |-0.7|, which is thus, the magnitude of -0.7, and is 0.7, the magnitude, that is, the absolute value of which is 0.7, too, of course.   That is to say that we have: |-0.7| = |0.7| = 0.7.

And in fact, doing subtractions, we can say that we do additions.   How come?

Subtracting a number, we can add the negative of the number.
For instance, we can have: 3 – 2 = 3 + (-2) = 1 + 2 + (-2) = 1 + 0 = 1.
And we can have: -5 – 3 = -5 + (-3) = -8.
So we have: 3 – 2 = 3 + (-2), and -5 – 3 = -5 + (-3).

So in general, we have: $A - B = A + (-B)$.

And we know that the negative of a negative is positive.
So for instance, we get: 4 – (-2) = 4 + 2 = 6, and -9 – (-4) = -9 + 4 = -5.

Thus in general, we have: $A - (-B) = A + B$.

So in sum, we can change the sign of a number multiplying (dividing) by –1.

And multiplying a number by –1, we get the negative of the number, and in that case, we add a negative sign to the number.

For instance, taking the negative of 2, we get: -1 x 2 = -2, which is thus, the negative of 2, and taking the negative of –3, we get: -1 x (-3) = -(-3) = 3, which is thus, the negative of -3.

And subtracting a number from another, we can add the negative of the number to the other number. That is to say that $A - B = A + (-B)$, and that $A - (-B) = A + B$.

And in the next section, we will move on to some basics of divisions, after doing several sets of examples in the next pages.

Examples 6

Find the sums below:

0. $4 + (-7) + 12 + (-8) + (-5)$

1. $-3 + (-8) + 9 + (-7) + (-2) + 5$

2. $-4 + (-3) + 5 + (-2) + 9 + (-7)$

3. $7 + (-27) + (-18) + (-26) + 23 + (-19)$

4. $-16 + (-27) + (-18) + 27 + 16 + (-17)$

5. $129 + (-26) + (-38) + (-12) + 28 + (-52)$

6. $-937 + 27 + (-18) + (-21) + 937 + (-27)$

7. $27 + (-39) + (-286) + (-9) + 785 + (-75) + (-23)$

8. $-334 + (-27) + 300 + 7 + (-389) + 9 + 30$

9. $467 + (-60) + 24 + (-7) + 300 + (-400) + (-300)$

A. $6789 + 247 + (-700) + (-47) + (-6000) + (-9)$

B. $-785 + 37 + (-428) + 80 + 300 + 17 + 400 + (-7)$

C. $4893 + (-3003) + 204 + (-90) + (-800) + 96$

D. $-5930 + 396 + (-56) + 4000 + (-300) + 1000 + 30 + (-300)$

34

Suggestions or Solutions
To the Examples 6

0. 4 + (-7) + 12 + (-8) + (-5)

$-7 - 5 = -12$

$4 + (-7) + 12 + (-8) + (-5)$

$= \underline{(-7) + 12 + (-5)} + 4 + (-8) = \underline{12 - 12} + 4 - 8 = \underline{0} + 4 - 8 = -4$

Of course, we can group together first, the positive numbers and negative numbers, and then, proceed with the work. There are many ways to get to the same destination.

What matters first in math is the accuracy. The next is the speed. We don't want to get the wrong solution fast, do we?
By the way, I make a lot of mistakes. So find the mistakes I made in the calculations below, and if any, correct them.

1. -3 + (-8) + 9 + (-7) + (-2) + 5

$9 + (-7) + (-2) = 9 + (-9) = 0$

$-3 + (-8) + 9 + (-7) + (-2) + 5 = -3 - 8 + 5 = -3 - 3 - 5 = 5 = -6$

2. -4 + (-3) + 5 + (-2) + 9 + (-7)

$-4 + (-3) + (-2) = -9$, and $(-2) + (-7) = -9$

$-4 + (-3) + 5 + (-2) + 9 + (-7) = 5 + (-7) = -2$, and we can get it the way below, too:
$-4 + (-3) + 5 + (-2) + 9 + (-7) = -4 + (-3) + 5 = -4 + 2 = -2.$

3. $7 + (-27) + (-18) + (-26) + 23 + (-19)$

$\{7 + (-27)\} + (-18) + \{(-26) + 23\} + (-19)$
$= -20 - 18 - 3 - 19$
$= -20 - 10 - 8 - 2 - 1 - 10 - 9$
$= -20 - 10 - 10 - 1 - 10 - 9$
$= -50 - 10$
$= -60$

4. $-16 + (-27) + (-18) + 27 + 16 + (-17)$

$\underline{-16} + (-27) + (-18) + 27 + \underline{16} + (-17)$
$= -18 - 17$
$= -18 - 12 - 5 = -30 - 5 = -35$

5. $129 + (-26) + (-38) + (-12) + 28 + (-52)$

$129 + (-26) + (-38) + (-12) + 28 + (-52)$
$= 129 - 26 - 10 - 12 - 52$
$= 103 - 10 - 12 - 52$
$= 101 - 10 - 10 - 52$
$= 81 - 52$ {Notice that $81 - 52 = 80 + 1 - 50 - 2 = 80 - 50 + 1 - 2$}
$= 30 - 1$
$= 29$

6. $-937 + 27 + (-18) + (-21) + 937 + (-27)$

$\underline{-937 + 27} + (-18) + (-21) + \underline{937 + (-27)}$
$= -18 - 21 = -18 - 2 - 19 = -39$
$= -9 - 9 - 21 = -9 - 30 = -39$

7. 27 + (-39) + (-286) + (-9) + 785 + (-75) + (-23)

27 + (-39) + (-286) + (-9) + 785 + (-75) + (-23)
= -4 – 39 – 286 – 9 – 75 + 785
= -43 – 200 – 80 – 6 – 9 – 75 + 700 + 80 + 5
= -43 + 500 – 10 – 75
= -43 – 85 + 500
= -128 + 100 + 400
= -28 + 100 + 300
= 378

8. -334 + (-27) + 300 + 7 + (-389) + 9 + 30

-334 + (-27) + 300 + 7 + (-389) + 9 + 30
= -34 – 20 – 380 + 30
= -54 – 350
= -404

9. 467 + (-60) + 24 + (-7) + 300 + (-400) + (-300)

467 + (-60) + 24 + (-7) + 300 + (-400) + (-300)
= 400 + 24 – 400
= 24

A. 6789 + 247 + (-700) + (-47) + (-6000) + (-9)

6789 + 247 + (-700) + (-47) + (-6000) + (-9)
= 89 – 247 –47 – 9
= 80 – 200
= -120

B.   **-785 + 37 + (-428) + 80 + 300 +17 + 400 + (-7)**

$\underline{-785}$ + 37 + (-428) + $\underline{80}$ + 300 +**17** + 400 + **(-7)**
= $\underline{-705}$ + 37 − 428 + 300 + **10** + 400
= -5 + 37 − 428 + 10
= 32 − 418
= 30 − 416   {Notice that 30 − 10 = 20.}
= 20 − 406
= -386

C.   **4893 + (-3003) + 204 + (-90) + (-800) + 96**

$\underline{4893}$ + $\underline{(-3003)}$ + 204 + $\underline{(-90)}$ + $\underline{(-800)}$ + 96
= $\underline{1000}$ + 204 + 96
= 1300

D.   **-5930 + 396 + (-56) + 4000 + (-300) + 1000 + 30 + (-300)**

$\underline{-5930}$ + **396** + (-56) + $\underline{4000}$ + **(-300)** + $\underline{1000}$ + $\underline{30}$ + (-300)
= $\underline{-900}$ + **96** − 56 − 300
= -1200 + 40
= -1160

Note again, there can be many ways to get the same.
So you don't have to do calculations the way above.

You might come up with the way it can be done better or the way you like, if you try, of course.

Tried but didn't work?

Well then, try again, if you want to, of course.
You never know until you try, and you know that.

Examples 7

Find the sums below:

0. -3 + (-6)

1. -5 + (-2)

2. -7 + (-4) + (-3) + (-5)

3. -6 + (-4) + (-6) + (-4) + (-2)

4. -1 + (-2) + (-3) + (-4) + (-9) + (-5) + (-7) + (-8) + (-6) + (-5)

5. -13 + (-17) + (-15) + (-17) + (-15) + (-15)

6. -16 + (-14) + (-19) + (-11) + (-18) + (-12)

7. -54 + (-46) + (-27) + (-63) + (-73) + (-12) + (-37) + (-88)

8. -123 + (-175) + (-164) + (-125) + (-136) + (-177)

9. -438 + (-357) + (-726) + (-562) + (-643) + (-274)

A. -1936 + (-730) + (-50) + (-14)+ (-270) + (-573)

B. -856 + (-700) + (-123) + (-70) + (-40) + (-7) + (-70) + (-74) + (-1260)

C. -492 + (-234) + (-10) + (-30) + (-538) + (-275) + (-498) + (-736) + (-462) + (-500) + (-225)

D. -12 + (-10) + (-234) + (-6) + (-428) + (-14) + (-80) + (-760) + (-8) + (-5) + (-10) + (-1865) + (-11) + (-9) + (-126) + (-50)

Suggestions or Solutions
To the Examples 7

Adding, you find the sum.

It is often the case you don't have to find it adding one at a time.

Sometimes, you can get the sum of all at once, too. It all depends on how you think, that is, how you do math.

0. -3 + (-6)

= 3 x (-1) + 6 x (-1)
= 9 x (-1)
= -9

1. -5 + (-2) = -7

2. -7 + (-4) + (-3) + (-5)

-7 + (-4) + (-3) + (-5)
= -10 + (-4) + (-5)
= -10 + (-9)
= -19

3. -6 + (-4) + (-6) + (-4) + (-2)

-6 + (-4) + (-6) + (-4) + (-2)
= -10 + (-10) + (-2)
= -22

4.   -1 + (-2) + (-3) + (-4) + (-9) + (-5) + (-7) + (-8) + (-6) + (-5)

$= -(1 + 2 + 3 + 4 + 5 + 6 + 7 + 8 + 9 + 5) = -(45 + 5) = -50$

Note that:

$1 + 2 + 3 + \ldots + 8 + 9 = (1 + 2 + 3 + \ldots + 8 + 9 + \mathbf{9} + \mathbf{8} + \ldots + \mathbf{2} + \mathbf{1}) \div 2$

$= \{(1 + \mathbf{9}) + (2 + \mathbf{8}) + (3 + \mathbf{7}) + \ldots + (8 + \mathbf{2}) + (9 + \mathbf{1})\} \div 2$

$= (10 \text{ x } 9) \div 2 = 90 \div 2 = 45.$ And we can put it the way below, too:

$1 + 2 + 3 + \ldots + 8 + 9 = 1 + 9 + 2 + 8 + 3 + 7 + 4 + 6 + 5 = 10 \text{ x } 4 + 5 = 45.$

And by the same token, we can get: $1 + 2 + 3 + \ldots + 100 = 101 \text{ x } 50 = 5050.$
What then, about this sum: $1 + 2 + 3 + \ldots + 1000$ or $2 + 4 + 6 + \ldots + 1000$?

One is: $1001 \text{ x } 500 = 500500$, and the other is: $2 \text{ x } 501 \text{ x } 250 = 250500.$

5.   -13 + (-17) + (-15) + (-17) + (-15) + (-15)

$= -(13 + 15 \text{ x } 3 + 17 \text{ x } 2)$
$= -(13 + \mathbf{45} + \mathbf{34})$
$= -(10 + 40 + 30 + \mathbf{3} + \mathbf{5} + \mathbf{4}) = -(80 + \mathbf{12}) = -92$

6.   -16 + (-14) + (-19) + (-11) + (-18) + (-12)

$\{-16 + (-14)\} + \{(-19) + (-11)\} + \{(-18) + (-12)\}$

$= -30 - 30 - 30 = -90$

7. **-54 + (-46) + (-27) + (-63) + (-73) + (-12) + (-37) + (-88)**

= {-54 + (-46)} + {(-27) + (-63)} + {(-73) + (-37)} + {(-12) + (-88)}
= -100 – 90 –110 – 100
= -400

8. **-123 + (-175) + (-164) + (-125) + (-136) + (-177)**

= {-123 + (-177)} + {(-175) + (-125)} + {(-164) + (-136)}
= -300 –300 – 300 = -900

9. **-438 + (-357) + (-726) + (-562) + (-643) + (-274)**

= {-438 + (-562)} + {(-357) + (-643)} + {(-726) + (-274)}
= -1000 – 1000 – 1000 = -3000

A. **-1936 + (-730) + (-50) + (-14) + (-270) + (-573)**

= {-1936 + (-50) + (-14)} + {(-730) + (-270)} + (-573)
= -2000 – 1000 – 573
= -3573

B. **-856 + (-700) + (-123) + (-70) + (-40) + (-7) + (-70) + (-74) + (-1260)**

{-856 + (-700)} + {(-123) + (-7)} + {(-70) + (-70) + (-74)} + {(-40) + (-1260)}
= -1556 – 130 – 214 – 1300
= -1560 – 130 – 210 – 1300
= -1560 – 340 – 1300   {Taking care of each digit}
= -2000 – 1100 – 100 = -3200

C.   **-492 + (-234) + (-10) + (-30) + (-538) + (-275)**

+ (-498) + (-736) + (-462) + (-500) + (-225)

$= -(492 + 538) - (234 + 736) - (275 + 225) - (498 + 462) - 500 - 40$

$= -(900 + 120 + 10) - 970 - 500 - (800 + 150 + 10) - 500 - 40$

$= -100 \times (9 + 1 + 9 + 5 + 8 + 1 + 5) + (-10) \times (2 + 1 + 7 + 5 + 1 + 4)$

$= -100 \times 38 - 10 \times 20$

$= -100 \times (38 + 2)$

$= -4000$

D.   **-12 + (-10) + (-234) + (-6) + (-428) + (-14) + (-80) + (-760) + (-8) + (-5) + (-10)**

+ (-1865) + (-11) + (-9) + (-126) + (-50)

$= \underline{-12 - 428} - 10 \underline{- 234 - 6} - 14 - 126 - 80 - 760 - 8 \underline{- 5 - 1865} - 10 - \underline{11 - 9} - 50$

$= \underline{-440} - 10 \underline{- 240} \underline{- 140} - 80 - 760 - 8 \underline{- 1870} - 10 \underline{- 20} - 50$

$= -1000 - 100 \times (4 + 2 + 1 + 7 + 8) - 10 \times (4 + 1 + 4 + 4 + 8 + 6 + 7 + 1 + 2 + 5) - 8$

$= -1000 - 100 \times 22 - 10 \times (4 \times 3 + 1 + 8 + 2 + 6 + 7 + 1 + 5) - 8$

$= -1000 - 2000 - 200 - 10 \times (12 + 1 + 10 + 13 + 6) - 8$

$= -3000 - 200 - 10 \times (30 + 12) - 8$

$= -3000 - 200 - 10 \times 42 - 8$

$= -3000 - 200 - 420 - 8$

$= -3628$

And of course, you don't have to do calculations the way above.

It's a good idea to come up with your examples, and do them as many ways as you can.

We don't learn much doing copy-and-paste, but learn a lot doing trial-and-error.
And in fact, we learn it doing it, and not just knowing it, if it's worth learning it.

And it seems knowledge doesn't do until it's in its action.
So if it is good and doable, don't just think about it.

Examples 8

Find the sums below:

0. $110 + (-19)$

1. $-71 + 114$

2. $51 + (-181)$

3. $-41 + 110$

4. $-11 + (-12) + (-13) + (-24) + (-19) + (-15) + (-17) + (-18) + (-26) + (-35)$

5. $-113 + (-117) + (-115) + (-117) + (-115) + (-115)$

6. $-116 + (-114) + (-119) + (-111) + (-118) + (-112)$

7. $-154 + (-246) + (-327) + (-463) + (-573) + (-612) + (-837) + (-988)$

8. $-1123 + (-1175) + (-1164) + (-1125) + (-1136) + (-1177)$

9. $-1438 + (-1357) + (-1726) + (-2562) + (-2643) + (-2274)$

A. $-1936 + (-730) + (-50) + (-14) + (-270) + (-573)$

B. $-8011 + 1103$

C.   -113 + (-117) + (-115) + (-117) + (-115) + (-115)

D.   -216 + (-214) + (-219) + (-311) + (-318) + (-312)

E.   -254 + (-246) + (-227) + (-363) + (-373) + (-412) + (-437) + (-488)

F.   -1123 + (-1175) + (-1164) + (-1025) + (-1036) + (-1077)

G.   -3438 + (-4357) + (-2726) + (-1562) + (-643) + (-2274)

H.   -1936 + (-1730) + (-150) + (-114)+ (-270) + (-573)

I.   -2856 + (-1700) + (-2123) + (-170) + (-240) + (-27) + (-270) + (-374) + (-1260)

J.   183 + (-127)

K.   -115 + 1105

L.   109 + (-1101)

M.   47 + (-107)

N.   -396 + 1397

O.   -967 + 2968

P.   1628 + (-529)

Q.   -1937 + 1839

R.   952 + (-42841)

Suggestions or Solutions
To the Examples 8

0.   110 + (-19)

$= 100 - 9$ {since $10 + (-10) = 0$}

$= 91$

1.   -71 + 114

$= 113 - 70$ {because $-1 + 1 = 0$}

$= 43$

2.   51 + (-181)

$= 50 - 180$ {since $1 + (-1) = 0$}

$= -130$

3.   -41 + 110

$= 70 - 1$ {because $-40 + 40 = 0$}

$= 69$

4.   -11 + (-12) + (-13) + (-24) + (-19) + (-15) + (-17) + (-18) + (-26) + (-35)

$= -11 + (-12) + (-13) + (-24) + (-19) + (-15) + (-17) + (-18) + (\underline{-26}) + (\underline{-35})$

$= -11 - 12 - 13 - \mathbf{14} - \mathbf{10} - 15 \underline{- 16 - 10} - 17 - 18 - 19 \underline{- \mathbf{20} - \mathbf{15}}$

$= -(11 + 12 + 13 + 14 + \ldots\ + 19 + 20 + \mathbf{10} + \mathbf{10} + 15)$

$= -\{(10 + 1) + (10 + 2) + (10 + 3) + \ldots + (10 + 9) + (10 + 10) + \mathbf{10} + \mathbf{10} + 15\}$

$= -(1 + 2 + 3 + \ldots + 9 + 10 + 10 \times 12 + 15)$

$= -(55 + 120 + 15) = -190$

How do we get though: $1 + 2 + 3 + \ldots + 9 + \mathbf{10} = \mathbf{55}$?

Suppose A = 1 + 2 + 3 + . . . + 9 + 10.   Then, we get:

2 x A = (1 + 2 + 3 + . . . + 9 + 10) + (1 + 2 + 3 + . . . + 9 + 10)

= (1 + 10) + (2 + 9) + (3 + 8) + . . . + (9 + 2) + (10 + 1) = 11 x 10 = 110.

So we get: 2 x A = 110 $\Rightarrow$ A = 55.

5.   -113 + (-117) + (-115) + (-117) + (-115) + (-115)

= -110 x 6 + {-(3 + 2 x 7 + 3 x 5)}
= -660 – (3 + 14 + 15) = -660 – 32 = -692

6.   -116 + (-114) + (-119) + (-111) + (-118) + (-112)

= -110 x 6 + {-(6 + 4 + 9 + 1 + 8 + 2)} = -660 – 30 = -690

7.   -154 + (-246) + (-327) + (-463) + (-573) + (-612) + (-837) + (-988)

= {-154 + (-246)} + {(-327) + (-463)} + {(-573) + (-837)} + {(-612) + (-988)}
= **-400** – 790 – 14<u>10</u> – **1600**
= **-2000** – 800 – 1400 = -4200

8.   -1123 + (-1175) + (-1164) + (-1125) + (-1136) + (-1177)

{We can see 6 of -1100s.}

= -1100 x 6 + {-(23 + 77 + 75 + 25 + 64 + 36)}
= -6600 – 300 = -6900

9.   -1438 + (-1357) + (-1726) + (-2562) + (-2643) + (-2274)

= **-1438** + (-2562) + (**-1357**) + (-2643) + (**-1726**) + (-2274)
= **-1500** − 2500 − **1400** − 2600 − **1800** − 2200
= -100 x (15 + 25 + 14 + 26 + 18 + 22)
= -100 x (40 + 40 + 40)
= -100 x 120 = -12000

A.   -1936 + (-730) + (-50) + (-14) + (-270) + (-573)

= {-1936 + (-14)} + {(-730) + (-270)} + (-50) + (-573)
= **-1950** − 1000 − **50** − 573
= **-2000** − 1000 − 573
= -3573

B.   -8011 + 1103

-8011 + <u>1103</u>
= **-8000** + <u>1000</u> + <u>100</u> − **11** + <u>3</u>
= -7000 + 100 − 8
= -6900 − 8
= -6908

C.   -113 + (-117) + (-115) + (-117) + (-115) + (-115)

= -113 + (-117) + (-11<u>5</u>) + (-11<u>7</u>) + (-11<u>5</u>) + (-11<u>5</u>)
= -110 x 6 − (**10** + <u>12</u> + **<u>10</u>**)
= -660 − 32
= -692

D. -216 + (-214) + (-219) + (-311) + (-318) + (-312)

{We can see three of (–200)s, three of (–300)s, and six of (–10)s.}

= -200 x 3 + (-300 x 3) + (-10 x 6) + {-(6 + 4 + 9 + 1 + 9 + 2)}
= -600 – 900 – 60 – 30
= -1590

E. -254 + (-246) + (-227) + (-363) + (-373) + (-412) + (-437) + (-488)

{We can see 3 of (–200)s, 2 of (–300)s, and 3 of (–400)s.}

= -600 – 600 – 1200 – (54 + 46 + 27 + 63 + 73 + 37 + 12 + 88)
= -2400 – (100 + 90 + 110 + 100)
= -2400 – 400 = -2800

F. -1123 + (-1175) + (-1164) + (-1025) + (-1036) + (-1077)

= -6000 – 300 – (23 + 77 + 75 + 25 + 64 + 36)
= -6300 – (100 + 100 + 100) = -6600

G. -3438 + (-4357) + (-2726) + (-1562) + (-643) + (-2274)

= -1000 x 12 + (-100) x (4 + 6 + 3 + 7 + 5 + 2)
+ (-10) x (3 + 7 + 5 + 2 + 6 + 4) – (8 + 2 + 7 + 3 + 6 + 4)

= -12000 – 100 x 27 – 10 x 27 – 30
= -12000 – 2700 – 270 – 30
= -14000 – 900 – 100
= -15000

H.   -1936 + (-1730) + (-150) + (-114)+ (-270) + (-573)

= **-1936** + (-150) + (-1730) + (-114)+ (-270) + (-573)

= **-2000** – 36 – 50 – 2000 –30 – 14 – 70 – 573

= -4000 – **86 – 100 – 14** – 573

= -4000 – **200** – 573 = -4773

I.   -2856 + (-1700) + (-2123) + (-170) + (-240) + (-27) + (-270) + (-374) + (-1260)

{We can see many of (–100)s.}

= -100 x (28 + 17 + 21 + 1 + 2 + 2 + 3 + 12) – (56 + 23 + 40 + 27 + 70 + 74 + 60)

= -100 x (28 + 2 + 17 + 1 + 2 + 21 + 3 + 12) – (**56 + 74** + 23 + 27 + 40 + 60 + 70)

= -100 x (30 + 20 + 36) – (**130** + 50 + 170)

= -100 x 86 – 350 = -8600 – 350 = -8950

J.   183 + (-127)   {We know: 100 + (-100) = 0}

= 83 – 27   {and we know: 20 + (-20) = 0}

= 63 – 7   {and we have: 7 + (-7) = 0, and 10 – 7 = 3}

= 53 + 3 = 56

K.   -115 + 1105   {We know: -105 + 105 = 0}

= -10 + 1000 = 990

L.   109 + (-1101)   {We know: 100 + (-100) = 0}

= 9 – 1001 = -1000 + 9 – 1 = -1000 + 8 = -992 – 8 + 8 = -992

M.   **47 + (-107)**   {7 + (-7) = 0}

= 40 − 100 = -60

N.   **-396 + 1397**   {-300 + 300 = 0}

= -96 + 1097   {-96 + 96 = 0}
= 1001

O.   **-967 + 2968**   {-967 + 967 = 0}
= 2001

P.   **1628 + (-529)**   {520 + (-520) = 0}

= 1108 − 9 = 1100 + 8 − 9 = 1100 − 1 = 1099

Q.   **-1937 + 1839**   {-1030 + 1030 = 0}

= -907 + 809   {-800 + 800 = 0}
= -107 + 9
= -100 − 7 + 9 = -100 + 2 = -98

R.   **952 + (-42841)**   {800 + (-800) = 0}

= 152 − 42041
= -4000 − **200 + 100** + 50 − 40 + 2 − 1
= -4000 − **100** + 10 + 1
= -4000 − 90 + 1 = -4000 − 89 = -4089

Examples 9

Do the subtractions below:

0. -1 – (-9)

1. -7 – (-4)

2. -15 – (-7)

3. -14 – (-14)

4. -81 – (-32)

5. -18 – (-15)

6. -15 – (-17)

7. -59 – (-72)

8. -74 – (-36)

9. -366 – (-307)

A. -907 – (-289)

B. -3628 – (-4779)

C. -1937 – (-849)

D. -572 – (-2301)

Suggestions or Solutions
To the Examples 9

Normally, we omit the + sign when we specify a positive number.

We have: **-A + B = B – A**, so we can get: **-A + A = A – A = 0**.

And we have: **-(-A) = A**, because we have: **-A = -1 x A**, so **-(-A) = -1 x (-1) x A = A**.

0. -1 – (-9)

We have: -(-9) = +9 = 9.   So we get: -1 – (-9) = -1 + 9 = -1 + 1 + 8 = 8

1. -7 – (-4) = -7 + 4 = -3 – 4 + 4 = -3

2. -15 – (-7) = -15 + 7 = -8 – 7 + 7 = -8

3. -14 – (-14) = -14 + 14 = 0

4. -81 – (-32)

= -81 + 32
= -80 – 1 + 30 + 2
= -50 – 30 + 30 – 1 + 2
= -50 – 1 + 1 + 1
= -50 + 1
= -49 – 1 + 1
= -49

5.   **-18 – (-15)** = -18 + 15 = -3 – 15 + 15 = -3

6.   **-15 – (-17)** = -15 + 17 = -15 + 15 + 2 = 2

7.   **-59 – (-72)**

= -59 + 72
= -50 – 9 + 70 + 2
= -50 + 50 + 20 – 7 – 2 + 2
= 20 – 7
= 13 + 7 – 7 = 13

8.   **-74 – (-36)**

= -74 + 36
= -70 – 4 + 30 + 6
= -40 – 30 + 30 – 4 + 4 + 2
= -40 + 2
= -38 – 2 + 2 = -38

9.   **-366 – (-307)**

= -366 + 307
= -300 – 60 – 8 + 300 + 7
= -60 – 1 – 7 + 7
= -61

A. -907 – (-289)

= -907 + 289
= -900 – 7 + 200 + 80 + 9
= -700 – 200 + 80 – 7 + 7 + 2
= -700 + 80 + 2
= -620 - 80 + 80 + 2
= -620 + 2
= -618 – 2 + 2 = -618

B. -3628 – (-4779)

= -3628 + 4779
= -3000 – 600 – 20 – 8 + 4000 + 700 + 70 + 9
= 1000 + 100 + 50 + 1 = 1151

C. -1937 – (-849)

= -1937 + 849
= -1000 – 900 – 30 – 7 + 800 + 40 + 9
= -1000 – 100 + 10 + 2
= -1000 – 90 + 2 = -1000 – 88 = -1088

D. -572 – (-2301)

= -572 + 2301
= -500 – 70 – 2 + 2000 + 300 + 1
= -200 – 70 – 1 + 2000
= 1800 – 70 – 1 = 1730 – 1 = 1729

Examples A

Do the subtractions below:

0. 7 – (-9)

1. 8 – (-12)

2. 5 – (-17)

3. 14 – (-14)

4. 38 – (-62)

5. 18 – (-82)

6. 15 – (-985)

7. 19 – (-481)

8. 47 – (-753)

9. 396 – (-204)

A. 967 – (-1032)

B. 628 – (-4372)

C. 1987 – (-1013)

D. 9572 – (-428)

58

Suggestions or Solutions
To the Examples A

Taking the negative of a number, we change the sign of it.
And changing the sign of a number, we multiply it by –1.

So taking the negative of a number, we multiply it by –1.

And multiplying it by –1, we add the minus sign to it.
So taking the negative of a number, we add the minus sign to it.
For instance, taking the negative of 3, we get: –3, called the negative 3.

So next, taking the negative of a negative, we add the minus sign to it, also.
For instance, taking the negative of –3, we get: –(–3), called the negative of negative 3.

And the negative of a negative is positive.
Thus, we get: –(–3) = +3, and we can omit the + sign if it's not needed.

So taking the negative of a negative, we get the positive of it.

And subtracting is minus, which means negative.
So subtracting a negative, we take the negative of the negative, and take the positive of it.
For instance, – (–5) = + 5, and we keep the + sign if it's needed.
And thus, subtracting a negative, we add its positive.

0. $7 - (-9) = 7 + 9 = 7 + 3 + 6 = 10 + 6 = 16$

1. $8 - (-12) = 8 + 12 = 8 + 2 + 10 = 10 + 10 = 20$

2. $5 - (-17) = 5 + 17 = 5 + 5 + 12 = 10 + 12 = 22$

3.   **14 – (-14)** = 14 + 14 = 28

4.   **38 – (-62)** = $\underline{3}$8 + $\underline{6}$2 = $\underline{90}$ + 10 = 100

5.   **18 – (-82)** = 1$\underline{8}$ + 8$\underline{2}$ = 90 + $\underline{10}$ = 100

6.   **15 – (-985)** = $\underline{1}$5 + 9$\underline{85}$ = 900 + $\underline{100}$ = 1000

7.   **19 – (-481)** = $\underline{1}$9 + 4$\underline{81}$ = 400 + $\underline{100}$ = 500

8.   **47 – (-753)** = 47 + 753 = 700 + 100 = 800

9.   **396 – (-204)** = 396 + 204 = 500 + 100 = 600

A.   **967 – (-1032)** = 9$\underline{67}$ + 10$\underline{32}$ = 1000 + 900 + $\underline{99}$ = 1999

B.   **628 – (-4372)** = $\underline{6}$28 + 4$\underline{3}$72 = 4000 + $\underline{900}$ + **100** = 5000

C.   **1987 – (-1013)** = 1987 + 1013 = 2000 + 900 + 100 = 3000

D.   **9572 – (-428)** = 9572 + 428 = 9000 + 900 + 100 = 10000

Examples B

Do the subtractions below:

0. $-1 - (-9)$ 1. $-7 - (-4)$ 2. $-5 - (-8)$

3. $-4 - (-14)$ 4. $-18 - (-28)$ 5. $-53 - (-23)$

6. $-15 - (-15)$ 7. $-19 - (-193)$ 8. $-47 - (-157)$

9. $-396 - (-256)$ A. $-765 - (-165)$ B. $-628 - (-528)$

C. $-4952 - (-902)$ D. $-9572 - (-9072)$

62

Suggestions or Solutions
To the Examples B

0. **-1 – (-9)**

= -1 + 9 = -1 + 1 + 8 = 8

1. **-7 – (-4)**

= -7 + 4 = -3 – 4 + 4 = -3

2. **-5 – (-8)**

= -5 + 8 = -5 + 5 + 3 = 3

3. **-4 – (-14)**

= -4 + 14 = -4 + 4 + 10 = 10

4. **-18 – (-28)**

= -18 + 28 = -18 + 18 + 10 = 10

5. **-53 – (-23)**

= -53 + 23 = -30 – 23 + 23 = -30

6.   **-15 – (-15)** = -15 + 15 = 0

7.   **-19 – (-193)**

= -19 + 193
= -10 – 9 + 100 + 90 + 3
= 100 + 80 – 6
= 100 + 74 + 6 – 6
= 174

8.   **-47 – (-157)**

= -47 + 157
= -40 – 7 + 100 + 50 + 7
= 100 + 10 = 110

9.   **-396 – (-256)**

= -396 + 256 = -300 – 90 – 6 + 200 + 50 + 6 = -100 – 40 = -140

A.   **-765 – (-165)**

= -765 + 165 = -700 – 65 + 100 + 65 = -600

B.   **-628 – (-528)**

= -628 + 528 = -600 – 28 + 500 + 28 = -100

C. **-4952 – (-902)**

= -4952 + 902 = -4000 – 902 – 50 + 902 = -4000 – 50 = -4050

D. **-9572 – (-9072)**

= -9572 + 9072 = -9072 – 500 + 9072 = -500

Examples C

Do the subtractions below:

0. -11 – (-19)   1. -27 – (-24)   2. -115 – (-17)

3. -114 – (-114)   4. -181 – (-132)   5. 1138 – (-162)

6. 118 – (-1182)   7. 1115 – (-985)   8. 119 – (-1481)

9. 147 – (-753)   A. 1396 – (-2204)   B. 2967 – (-1032)

C. 1628 – (-4372)   D. -18 – (-108)   E. -15 – (-175)

F. -579 – (-72)   G. -734 – (-36)   H. -153 – (-104)

I. -15 – (-105)   J. -197 – (-193)   K. -443 – (-157)

L. -1396 – (-256)   M. -1765 – (-165)   N. -628 – (-1528)

O. -4952 – (-5902) P. -9572 – (-9073) Q. -368 – (-1307)

R. -907 – (-1986)   S. -727 – (-1726)   T. -1937 – (-947)

U. -3201 – (-2301)

Suggestions or Solutions
To the Examples C

0. **-11 – (-19)** = -11 + 19 = -11 + 11 + 8 = 8

1. **-27 – (-24)** = -27 + 24 = -24 – 3 + 24 = -3

2. **-115 – (-17)** = -115 + 17 = -100 – 15 + 15 + 2 = -100 + 2 = -98

3. **-114 – (-114)** = -114 + 114 = 0

4. **-181 – (-132)** = -181 + 132 = -100 – 80 – 1 + 100 + 30 + 2 = -50 + 1 = -49

5. **1138 – (-162)** = 1138 + 162 = 1000 + 200 + 90 + 10 = 1000 + 300 = 1300

6. **118 – (-1182)** = 118 + 1182 = 1000 + 200 + 90 +10 = 1300

7. **1115 – (-985)** = 1115 + 985 = 1000 + 1000 + 90 + 10 = 2100

8. **119 – (-1481)** = 119 + 1481 = 500 + 90 + 10 + 1000 = 1600

9. **147 – (-753)** = 147 + 753 = 800 + 90 + 10 = 900

A.   **1396 – (-2204)** = 1396 + 2204 = 3000 + 500 + 100 = 3600

B.   **2967 – (-1032)** = 2967 + 1032 = 3000 + 900 + 99 = 3999

C.   **1628 – (-4372)** = 1628 + 4372 = 5000 + 900 + 90 + 10 = 6000

D.   **-18 – (-108)** = -18 + 108 = -10 – 8 + 100 + 8 = 90

E.   **-15 – (-175)** = -15 + 175 = -15 + 15 + 160 = 160

F.   **-579 – (-72)** = -579 + 72 = -500 – 72 – 7 + 72 = -507

G.   **-734 – (-36)** = -734 + 36 = -700 – 34 + 34 + 2 = -700 + 2 = -698

H.   **-153 – (-104)** = -153 + 104 = -100 – 50 – 3 + 100 + 4 = -50 + 1 = -49

I.   **-15 – (-105)** = -15 + 105 = -10 + 100 = 90

J.   **-197 – (-193)** = -197 + 193 = -7 + 3 = -4

K.   **-443 – (-157)**

= -443 + 157 = -400 – 40 – 3 + 100 + 50 + 7 = -300 + 10 + 4 = -290 + 4 = -286

L.   **-1396 – (-256)** = -1396 + 256 = -1000 – 100 – 40 = -1140

M.   **-1765 – (-165)** = -1765 + 165 = -1000 – 600 = -1600

N.   **-628 – (-1528)** = -628 + 1528 = 1000 – 100 = 900

O.   **-4952 – (-5902)** = -4952 + 5902 = 1000 – 50 = 950

P.   **-9572 – (-9073)** = -9572 + 9073 = -500 + 1 = -499

Q.   **-368 – (-1307)** = -368 + 1307 = 1000 – 60 – 1 = 940 – 1 = 939

R.   **-907 – (-1986)** = -907 + 1986 = 1000 + 80 – 1 = 1079

S.   **-727 – (-1726)** = -727 + 1726 = 1000 – 1 = 999

T.   **-1937 – (-947)** = -1937 + 947 = -1000 + 10 = 990

U.   **-3201 – (-2301)** = -3201 + 2301 = -1000 + 100 = -900

Examples D

Find the products below:

0. 12 x (-1)

1. -1 x (-35)

2. 3 x (-4)

3. -5 x 7

4. -12 x 2

5. 21 x (-3)

6. 17 x (-5)

7. 12 x (-12)

8. -15 x 15

9. -25 x 25

A. -35 x 35

B. 45 x (-45)

C. 55 x (-55)

D. 65 x (-65)

E. 75 x (-75)

F. -85 x 85

G. 95 x (-95)

Suggestions or Solutions
To the Examples D

0. 12 x (-1)

Multiplying a nonzero number by –1, we just change the sign of the number.
So we get: 12 x (-1) = -12.   And for the same reason, we can get: -95 x (-1) = 95.

So multiplying a positive number by –1, we get the negative number with the same
magnitude, and also, multiplying a negative number by –1, we get the positive number
with the same magnitude, too.

Note that 0 is neither positive nor negative, and thus, has no sign.
And multiplying a number by 0, or multiplying 0 by a number, we get 0.

1. (-1) x (-35)

Multiplying –1 by a nonzero number, we change the sign of the number, too.
And we have: **A x B = B x A**.   So we get: (-1) x (-35) = (-35) x (-1) = 35.

And for the same reason, we can get: -1 x 17 = -17.

And of course, diving a nonzero number by –1, we change the sign of the number, too.
So for instance, we get: 3/(-1) = -3, and (-5)/(-1) = 5.

It is <u>usually</u> the case however, dividing –1 by a nonzero number, we not only change the
sign of the number, but change the magnitude of it, too.
For instance, we get: -1 ÷ 3 = -1/3, and -1 ÷ 3 ≠ -3.   What then, is the exception?

We have two cases. In one, the nonzero number is 1, and in the other, it is –1.
So we get: -1 ÷ **1** = -1/1 = **-1**, and we get: -1 ÷ **(-1)** = -1/(-1) = **1**.

2.   3 x (-4)

We have: **-A = -1 x A**, and we have: **A x B= B x A**.

So we get: 3 x (-4) = 3 x (-1) x 4 = -1 x 3 x 4 = -1 x 12 = -12.

And we can put it the way below, too:

   Adding together three of (-4)s, we get -12, and can put it this way: 3 x (-4).
   So we get: 3 x (-4) = -12.

And we know multiplying a nonzero by –1, we change the sign.
So in multiplication, if we have <u>an odd number of negatives</u>, the product is <u>negative</u>.
And if we have an <u>even number of negatives</u>, the product is <u>positive</u>.

So in general, taking the product of numbers positive and negative mixed together, we can just take all the numbers as positive numbers, take the product, and then, add the negative sign to the product if odd is the number of the negatives.

So for instance, doing -2 x (-3) x (-4), we do this first: 2 x 3 x 4, and get 24, and then, add the negative sign to 24, that is, we get –24, because in the entire set of the multiplications, the number of negatives is 3, which is odd.

That is to say that we get: -2 x (-3) x (-4) = -(2 x 3 x 4) = -24.
And we get: -2 x 3 x (-4) = 2 x 3 x 4 = 24.

3.   -5 x 7

We have: 5 x 7 = 35, so we get: -5 x 7 = -35.
And we have a rule where **A x B= B x A**.

So in fact, we have: -5 x 7 = 7 x (-5), which means the sum of 7 of (-5)s, which means that we get: -5 + (-5) + (-5) + (-5) + (-5) + (-5) + (-5) = -35.

4.   -12 x 2

We have: 12 x 2 = 24, so we get: -12 x 2 = -24, since it has one negative, that is, the number of the negatives is odd.

5.   21 x (-3)

We have: 21 x 3 = 63, so we get: 21 x (-3) = -63, which equals: -21 x 3, also.

6.   17 x (-5)

We can have: 17 x 5 = (10 + 7) x 5 = 10 x 5 + 7 x 5 = 50 + 35 = 85
So we get: 17 x (-5) = -17 x 5 = -85

7.   12 x (-12)

We can have: 12 x 12 = 12 x (10 + 2) = 12 x 10 + 12 x 2 = 120 + 24 = 144.
So we get: 12 x (-12) = -12 x 12 = -144

8.   -15 x 15

15 x 15 = 15 x (10 + 5) = 15 x 10 + 15 x 5 = 150 + (10 + 5) x 5 = 150 + 50 + 25 = 225
So we get: -15 x 15 = 15 x (-15) = -225

9.   -25 x 25

25 x 25 = (20 + 5) x 25 = 20 x 25 + 5 x 25 = 500 + 5 x (20 + 5) = 500 + 100 + 25 = 625
So we get: -25 x 25 = 25 x (-25) = -625

A.   **-35 x 35**

We can have: 35 x 35 = (30 + 5) x 35 = 30 x 35 + 5 x 35
= **30 x (30 + 5)** + <u>5 x (30 + 5)</u> = **900 + 150** + <u>150 + 25</u> = 900 + 300 + 25 = 1225

So we get: -35 x 35 = 35 x (-35) = -1225

B.   **45 x (-45)**

We can have: 45 x 45 = 45 x (40 + 5) = 45 x (2 x 20 + 5)
= 90 x 20 + 45 x 5 = 1800 + (40 + 5) x 5 = 1800 + 200 + 25 = 2025

So we get: 45 x (-45) = -45 x 45 = -2025

C.   **55 x (-55)**

55 x 55 = 55 x (50 + 5) = 55 x 50 + 55 x 5

55 x 5 = (50 + 5) x 5 = 250 + 25 = 275

55 x 50 = 55 x 5 x 10 = 275 x 10 = 2750

55 x 50 + 55 x 5 = 2<u>75</u>0 + <u>2</u>75 = 2000 + <u>9</u>00 + **120** + 5 = 2000 + 1020 + 5 = 3025

So we get: 55 x (-55) = -55 x 55 = -3025

D.   **65 x (-65)**

65 x 65 = 65 x (60 + 5) = 65 x 60 + 65 x 5 = (60 + 5) x 60 + (60 + 5) x 5
= 60 x 60 + 5 x 60 + 60 x 5 + 5 x 5 = 60 x 60 + 2 x 5 x 60 + 5 x 5 = 3600 + 600 + 25
= 3000 + 1200 + 25 = 4225

So we get: 65 x (-65) = -65 x 65 = -4225.

E.   75 x (-75)

75 x 75 = (70 + 5) x (70 + 5) = 70 x 70 + 2 x 5 x 70 + 5 x 5 = 4900 + 700 + 25

= 4000 + 1600 + 25 = 5625

So we get: 75 x (-75) = -75 x 75 = -5625

F.   -85 x 85

85 x 85 = (80 + 5) x (80 + 5) = 80 x 80 + 2 x 5 x 80 + 5 x 5 = 6400 + 800 + 25
= 6000 + 1200 + 25 = 7225

So we get: -85 x 85 = 85 x (-85) = -7225

G.   95 x (-95)

95 x 95 = 90 x 90 + 2 x 5 x 90 + 5 x 5 = 8100 + 900 + 25 = 8000 + 1000 + 25 = 9025

So we get: 95 x (-95) = -95 x 95 = -9025

And again, the way calculations done in the examples above is just one of many ways. So you don't have to do calculations the way above.

Examples E

Find the products below:

0. -12 x (-12)

1. -11 x (-11)

2. -101 x (-11)

3. -11 x (-101)

4. -1001 x (-25)

5. -1001 x (-125)

6. -10001 x (-37)

7. -10001 x (-937)

8. -10001 x (-5937)

9. -100001 x (-12345)

A. -100001 x (-1234)

B. -100001 x (-123)

C. -2002 x (-123)

D. -202 x (-12)

E. -20002 x (-123)

F. -3003 x (-111)

G. -3003 x (-222)

H. -3003 x (-123)

I. -3003 x (-125)

J. -303 x (-125)

K. -303 x (-111)

Suggestions or Solutions
To the Examples E

Taking the product of numbers positive and negative mixed together, we can just take all the numbers as positive numbers, take the product, and then, put the negative sign if odd is the number of the negatives.

So for instance, we get: -2 x (-3) x (-4) = -24, and we get: -2 x 3 x (-4) = 24.

0. -12 x (-12)
{It has two negatives, that is, an even number of negatives. So the product is positive.}

So we get: -12 x (-12) = 12 x 12 = (10 + 2) x 12 = 10 x 12 + 2 x 12 = 120 + 24 = 144.

1. -11 x (-11)
{It has two negatives, that is, an even number of negatives. So the product is positive.}

So we get: -11 x (-11) = 11 x 11 = (10 + 1) x 11 = 10 x 11 + 1 x 11 = 110 + 11 = 121

2. -101 x (-11)

= 101 x 11 = 101 x (10 + 1) = 101 x 10 + 101 x 1 = 1010 + 101 = 1000 + 100 + 10 + 1 = 1111

3. -11 x (-101) = 11 x 101 = 101 x 11 = 1111

4. -1001 x (-25) = 1001 x 25 = (1000 + 1) x 25 = 25000 + 25 = 25025

5.   **-1001 x (-125)** = 1001 x 125 = (1000 + 1) x 125 = 125000 + 125 = 125125

6.   **-10001 x (-37)** = 10001 x 37 = (10000 + 1) x 37 = 370000 + 37 = 370037

7.   **-10001 x (-937)** = 10001 x 937 = (10000 + 1) x 937 = 9370000 + 937 = 9370937

8.   **-10001 x (-5937)** = 10001 x 5937 = 59370000 + 5937 = 59375937

9.   **-100001 x (-12345)** = 100001 x 12345 = 1234500000 + 12345 = 1234512345

A.   **-100001 x (-1234)** = 100001 x 1234 = 123400000 + 1234 = 123401234

B.   **-100001 x (-123)** = 100001 x 123 = 12300123

C.   **-2002 x (-123)** = 2002 x 123 = 2 x 1001 x 123 = 2 x 123123 = 246246

D.   **-202 x (-12)** = 202 x 12 = 2 x 101 x 12 = 2 x 1212 = 2424

E.   **-20002 x (-123)** = 20002 x 123 = 2 x 10001 x 123 = 2 x 1230123 = 2460246

F.   **-3003 x (-111)** = 3003 x 111 = 3 x 1001 x 111 = 3 x 111111 = 333333

G. -3003 x (-222)

= 3003 x 222 = 3 x 1001 x 111 x 2 = 3 x 111111 x 2 = 333333 x 2 = 666666

H. -3003 x (-123) = 3003 x 123 = 3 x 1001 x 123 = 3 x 123123 = 369369

I. -3003 x (-125)

= 3003 x 125
= 3 x 1001 x 125
= 3 x 125125
= 3 x (125 x 1000 + 125)
= 3 x **125** x 1000 + 3 x 125
= (**300** + 60 + **15**) x 1000 + 300 + 60 + 15
= 375 x 1000 + 375 = 375375

J. -303 x (-125)

= 3 x 101 x 125
= 3 x (100 + 1) x 125
= 3 x (12500 + 125)
= 3 x **125** x 100 + 3 x 125
= (**300** + 60 + **15**) x 100 + 300 + 60 + 15
= 375 x 100 + 375
= (375 + 3) x 100 + 75 = 37800 + 75 = 37875

K. -303 x (-111)

= 3 x 101 x 111 = 3 x (100 + 1) x 111 = 3 x (11100 + 111) = 3 x 11211 = 33633

Examples F

Find the products below:

0. 3 x (-1) x (-1)

1. 4 x (-3) x (-5) x 2 x (-7)

2. -5 x (-6) x (-1) x (-1) x (-1) x 3

3. 2 x (-2) x (2) x (-2)

4. 3 x (-3) x (-3) x (-3)

5. 4 x (-4) x (-2) x (-1) x (-1)

6. (-1) x (-1) x (-2) x (-1) x (-1)

7. (-7) x (-21) x (-12) x 0

8. (-1) x (-2) x (-3) x (-4)

9. (-1) x (-2) x (-3) x (-4) x (-1)

A. (-10) x 10 x (-10) x 10 x (-10) x (-1)

B. 10 x 10 x 10 x 10 x (-1) x (-10) x 10

C. -10 x (-10) x (-10) x (-10) x (-10) x (-1)

D. -10 x (-10) x (-10) x (-10) x (-10) x (-1) x (-5)

E. 10 x (-10) x (-10) x 10 x (-11) x (-1) x (-5)

F. 11 x (-101) x 2 x (-100001) x (-1) x (-3)

Suggestions or Solutions
To the Examples F

Taking the product of numbers positive and negative mixed together, we can just take all the numbers as positive numbers, take the product, and then, put the negative sign if odd is the number of the negatives.

So for instance, we get: -2 x (-3) x (-4) x (-1) = 24, and we get: -2 x 3 x (-4) x (-1) = -24.

0.   3 x (-1) x (-1)   It has two negatives, so the product is positive.

So we get: 3 x (-1) x (-1) = 3 x 1 x 1 = 3 x 1 = 3

1.   4 x (-3) x (-5) x 2 x (-7)   It has three negatives, so the product is negative.

So we can just begin with 4 x 3 x 5 x 2 x 7.   Then, we get:
4 x 3 x 5 x 2 x 7 = 12 x 10 x 7 = 12 x 70 = (10 + 2) x 70 = 700 + 140 = 840.

And of course, we can put it the way below, too:

4 x (-3) x (-5) x 2 x (-7) = -(**4 x 3** x 5 x 2 x 7) = -(**12** x 10 x 7)
= -(12 x 70) = -{(**10** + 2) x 70} = -(**700** + 140) = -840

2.   -5 x (-6) x (-1) x (-1) x (-1) x 3   It has 5 negatives, so the product is negative.

And getting the magnitude of the product, we get:
5 x 6 x 1 x 1 x 1 x 3 = 5 x 6 x 3 = 30 x 3 = 90.   So the product is –90.

3.   2 x (-2) x (2) x (-2)   It has two negatives, so the product is positive.
So we can just take: 2 x 2 x 2 x 2 = 4 x 4 = 16

4.   **3 x (-3) x (-3) x (-3)**   It has three negatives, so the product is negative.
So taking first, the magnitude of the product, we get: 3 x 3 x 3 x 3 = 9 x 9 = 81.
Therefore, the product is –81.

5.   **4 x (-4) x (-2) x (-1) x (-1)** = 4 x 4 x 2 = 4 x 8 = 32

6.   **(-1) x (-1) x (-2) x (-1) x (-1)** = -2

7.   **(-7) x (-21) x (-12) x 0** = 0

8.   **(-1) x (-2) x (-3) x (-4)** = 1 x 2 x 3 x 4 = 2 x 3 x 4 = 6 x 4 = 24

9.   **(-1) x (-2) x (-3) x (-4) x (-1)** = -2 x 3 x 4 = -6 x 4 = -24

A.   **(-10) x 10 x (-10) x 10 x (-10) x (-1)** = 10 x 10 x 10 x 10 x 10 = 100,000

B.   **10 x 10 x 10 x 10 x (-1) x (-10) x 10** = 1,000,000

C.   **-10 x (-10) x (-10) x (-10) x (-10) x (-1)** = 100,000

D.   **-10 x (-10) x (-10) x (-10) x (-10) x (-1) x (-5)** = -500,000

E.   **10 x (-10) x (-10) x 10 x (-11) x (-1) x (-5)** = -550000

82

F.   11 x (-101) x 2 x (-100001) x (-1) x (-3)

{It has four negatives, so the product is positive.}

= 11 x 101 x 2 x 100001 x 3
= 11 x 101 x 100001 x 2 x 3
= <u>11 x 101</u> x 100001 x 6
= 1111 x 100001 x 6
= 111101111 x 6
= 666606666

3. The Basics 3

Doing arithmetic, we do operations called additions, subtractions, multiplications, and divisions. It can be said however, doing arithmetic, we do either of two operations, one is addition, and the other is multiplication.   How come?

Subtracting a number, we can add the negative of the number.

For instance, we can have: $2 - 5 = 2 + (-5) = 2 + (-2) + (-3) = 0 + (-3) = -3$.
And we can have: $-7 - 9 = -7 + (-9) = -16$.
So we have: $2 - 5 = 2 + (-5)$, and $-7 - 9 = -7 + (-9)$.
So in general, we have: $A - B = A + (-B)$.

And we know that the negative of a negative is positive.
So for instance, we get: $4 - (-2) = 4 + 2 = 6$, and $-9 - (-4) = -9 + 4 = -5$.
Thus in general, we have: $A - (-B) = A + B$.

What then, about divisions?

Dividing by a number, we can multiply by the reciprocal of the number.

For instance, dividing 6 by 3, we can multiply 6 by 1/3, since 1/3 is the reciprocal of 3. Then, we get 2, which is the quotient we get dividing 6 by 3. And we can take 2 as a product, too, because it is the product we get multiplying 6 by 1/3.

And dividing 8 by -2, we can multiply 8 by -1/2, since -1/2 is the reciprocal of -2. Then, we get -4, which is the quotient we get dividing 8 by -2. And we can take -4 as a product, too, because it is the product we get multiplying 8 by -1/2.

For another instance, dividing 6/5 by 2/5, we can multiply 6/5 by 5/2, since 5/2 is the reciprocal of 2/5. Then, we get 3, which is the quotient we get dividing 6/5 by 2/5. And we can take 3 as a product, too, because it is the product we get multiplying 6/5 by 5/2.

And dividing -9/5 by -3/5, we can multiply -9/5 by -5/3, because -5/3 is the reciprocal of -3/5. Then, we get 3, which is the quotient we get dividing -9/5 by -3/5. And we can take 3 as a product, too, because it is the product we get multiplying -9/5 by -5/3.

What then, is a reciprocal?

Dividing 1 by a number nonzero, we get the reciprocal of the number.

So for instance, 1/3 is the reciprocal of 3, because dividing 1 by 3, we get 1/3.
And -1/5 is the reciprocal of -5, because dividing 1 by -5, we get -1/5.
A zero has no reciprocal, because we have no division by 0.

And multiplying a number by its reciprocal, we get 1. And if we get 1 multiplying two numbers by each other, the two are said to be reciprocal of each other.

For instance, the reciprocal of 2/3 is 3/2, because we get: $\dfrac{2}{3} \cdot \dfrac{3}{2} = \dfrac{6}{6} = 1$.

And the reciprocal of -5/4 is -4/5, and we get: $-(5/4) \cdot -(4/5) = 20/20 = 1$.

And thus, if the product of two numbers is 1, the two are reciprocal of each other.

For instance, 7/5 is the reciprocal of 5/7, which is the reciprocal of 7/5, too, because we get: $(7/5) \cdot (5/7) = 1$. And -4/3 is the reciprocal of -3/4, which is the reciprocal of -4/3, also, because we get: $-(4/3) \cdot -(3/4) = 1$.

So in general, if $AB = 1$, A is the reciprocal of B, and B is the reciprocal of A, too, and we say that A and B are reciprocal of each other.

And if A and B both are not zero, $\dfrac{A}{B}$ and $\dfrac{B}{A}$ are reciprocal of each other.

So $\dfrac{A}{B}$ is the reciprocal of $\dfrac{B}{A}$, which is the reciprocal of $\dfrac{A}{B}$, too.

Swapping thus, the numerator and the denominator, we get the reciprocal.
What then, about the reciprocal of 5?

We know: 5 = 5/1, so taking the reciprocal of 5, we get: 1/5.
What then, about the reciprocal of 1?

The reciprocal of 1 is 1, because 1 x 1 = 1, and if the product of two numbers is 1, the two are reciprocal of each other, that is, one is the reciprocal of the other, and vice versa.

And dividing by a number, we can multiply by the reciprocal of the number.

Dividing for instance, 6 by 3, we can multiply 6 by 1/3, and get 2.

And dividing 1 by 1/4, we can multiply 1 by 4, and get 4.
In fact, adding together four of (1/4)s, we get 1, that is, multiplying 4 by 1/4, we get 1.

Next, we know if in multiplication, the number of negatives is odd, the product is negative. And if the number of negatives is even, the product is positive.

The same is true for divisions, too.

So in division, if the <u>number of negatives is odd</u>, the quotient is <u>negative</u>.
And if the number of negatives is even, the quotient is positive.

It's because a division by a number is a multiplication by its reciprocal, and taking a reciprocal of a number, we don't change the sign of the number.

So for instance, we get: 8 ÷ (-2) = -4, that is, 8/(-2) = -4.   So we get: 8/(-2) = -8/2.

In fact, multiplying by the same number, the numerator and the denominator, we get the same fraction. So in the fraction 8/(-2), multiplying 8 and -2 both by –1, we get -8/2.

And for another instance, 24 ÷ (-4) ÷ (-2) = 3, that is, 24/(-4)/(-2) = 3.
That's because: 24/(-4) = -6.   So we get: 24/(-4)/(-2) = -6/(-2) = 3.

And we can put it the way below, too:

24/(-4)/(-2) = 24/8 = 3, because 24/(-4)/(-2) = 24/{(-4)·(-2)} = 24/8 = 3.

And for another instance, we get: -24 ÷ (-4) ÷ (-2) = -3, that is, -24/(-4)/(-2) = -3.

That's because: -24/(-4) = 6.   So we get: -24/(-4)/(-2) = 6/(-2) = -3.

And we can put it the way below, too:

-24/(-4)/(-2) = -24/8 = -3, because -24/(-4)/(-2) = -24/{(-4)·(-2)} = -24/8 = -3.

And for another instance, we get: $\dfrac{-2}{3}\cdot\dfrac{5}{-7}\cdot\dfrac{-8}{9}\cdot\dfrac{-12}{11}\cdot\dfrac{8}{-3} = -\dfrac{2}{3}\cdot\dfrac{5}{7}\cdot\dfrac{8}{9}\cdot\dfrac{12}{11}\cdot\dfrac{8}{3}$

That's because the entire operation has an odd number of negatives.

And for another instance, we can get:

-37/(-7)/5/(-2) = -37/70, and -37/(-7)/(-5)/(-2) = 37/70.

So in general, doing divisions of numbers positive and negative mixed together, we can just take all the numbers as positive numbers, take the quotient, and then, add the negative sign to the quotient if odd is the number of all the negatives.

For instance, doing this: -37/(-7)/5/(-2), do this: -37/7/5/2, because the number of negatives is 3, which is odd.

For another instance, doing this: -37/(-7)/(-5)/(-2), just do this: 37/7/5/2, because the number of negatives is 4, which is even.

Examples G

Do the divisions below:

0. $-6 \div 3 = -6 / 3 = \dfrac{-6}{3} = -\dfrac{6}{3} =$

1. 1 / (-1)

2. -7 / 7

3. 12 / (-3)

4. 121 / (-11)

5. 225 / (-25)

6. -625 / 25

7. -1225 / 35

8. -2025 / 45

9. 3025 / -55

A. 9025 / -95

B. 5625 / -75

C. 7225 / -85

D. -4225 / 65

E. 123123 / -1001

F. 1230123 / -10001

G. -12300123 / 100001

Suggestions or Solutions
To the Examples G

In division, if the <u>number of negatives is odd</u>, the quotient is <u>negative</u>.
And if the number of negatives is even, the quotient is positive.   So for instance:

$8 \div (-2) = -4$, that is, $8/(-2) = -4$.   So we get: $8/(-2) = -8/2$.

$24 \div (-4) \div (-2) = 3$, that is, $24/(-4)/(-2) = 3$.

$-24 \div (-4) \div (-2) = -3$, that is, $-24/(-4)/(-2) = -3$.

$-37/(-7)/5/(-2) = -37/70$, and $-37/(-7)/(-5)/(-2) = 37/70$.

So in general, doing divisions of numbers positive and negative mixed together, we can just take all the numbers as positive numbers, take the quotient, and then, add the negative sign to the quotient if odd is the number of all the negatives.

For instance, doing this: $-17/(-9)/5/(-2)$, do this: $-17/9/5/2$, because the number of negatives is 3, which is odd.

And doing this: $-28/(-5)/(-3)/(-2)$, just do this: $28/5/3/2$, because the number of negatives is 4, which is even.

0. $-6 \div 3 = -6 / 3 = \dfrac{-6}{3} = -\dfrac{6}{3} = -2$ {The number of negatives is odd in the division.}

And we can put it the way below, too:

$\textbf{-6 / 3} = \dfrac{-6}{3} = \dfrac{-1 \times 6}{3} = (-1) \times \dfrac{6}{3} = -1 \times 2 = -2.$

1.   **1 / (-1) =** $\dfrac{1}{-1} = -\dfrac{1}{1} = -1$.   {The number of negatives is odd in the division.}

And we can put it this way, too: $1 / (-1) = \dfrac{1}{-1} = \dfrac{1 \cdot (-1)}{-1 \cdot (-1)} = \dfrac{-1}{1} = -1$.

2.   **-7 / 7 =** $\dfrac{-7}{7} = -\dfrac{7}{7} = -1$.   {The number of negatives is odd in the division.}

We can put it this way, too: $\dfrac{-7}{7} = \dfrac{-1 \times 7}{7} = -1 \times \dfrac{7}{7} = -1 \times 1 = -1$

3.   **12 / (-3) =** $\dfrac{12}{-3} = -\dfrac{12}{3} = -4$.

4.   **121 / (-11) =** $\dfrac{121}{-11} = -\dfrac{121}{11} = -\dfrac{11 \times 11}{11} = -11$

5.   **225 / (-25) =** $\dfrac{225}{-25} = -\dfrac{225}{25} = -\dfrac{9 \times 25}{25} = -9$

6.   **-625 / 25 =** $\dfrac{-625}{25} = -\dfrac{625}{25} = -\dfrac{25 \times 25}{25} = -25$

7.   **-1225 / 35**

$1225 = 1200 + 25 = 12 \times 100 + 25 = 12 \times 5 \times 20 + 25 = 12 \times 5 \times 5 \times 4 + 25$
$= 12 \times 4 \times 25 + 25 = 48 \times 25 + 25 = (48 + 1) \times 25 = 49 \times 25$.

And also, we can put 1225 the way below, too:
$1225 = 49 \times 5 \times 5 = 7 \times 7 \times 5 \times 5 = 7 \times 5 \times 7 \times 5 = 35 \times 35$.

Therefore, $1225 = 35 \times 35$, $1225/35 = 35$, and $-1225/35 = -35$.
Note however, you don't have to do calculations the way above.
There can be many ways to calculate.

8.   -2025 / 45

2025 = 2000 + 25 = 5 x 400 + 25 = 5 x 5 x 80 + 25 = 25 x 80 + 25 = 25 x (80 + 1)
= 25 x 81 = 5 x 5 x 9 x 9 = 5 x 9 x 5 x 9 = 45 x 45.

So we get: 2025/45 = 45, and thus, we get: -2025/45 = -45.

9.   3025 / -55

3025 = 3000 + 25 = 5 x 600 + 25 = 5 x 6 x 100 + 25 = 5 x 6 x 5 x 20 + 25
= 5 x 5 x 6 x 20 + 25 = 25 x 120 + 25 = 25 x (120 + 1) = 25 x 121

121 = 11 x 11
25 x 121 = 25 x 11 x 11 = 5 x 5 x 11 x 11 = 5 x 11 x 5 x 11 = 55 x 55

So we get: 3025/55 = 55, and thus, we get: 3025/(-55) = -55.

A.   9025 / -95

We have: 9025 = 9000 + 25
Meanwhile: 9000 = 90 x 100 = 90 x 5 x 20 = 90 x 5 x 5 x 4 = 25 x 90 x 4 = 25 x 360
So we get: 9025 = 25 x 360 + 25 = 25 x (360 + 1) = 25 x 361

Next, we can get: 95 = 90 + 5 = 9 x 2 x 5 + 5 = 18 x 5 + 5 = (18 + 1) x 5 = 19 x 5

And we have: 361 = 19 x 19.   So we get:

9025 = 25 x 19 x 19 = 5 x 5 x 19 x 19 =5 x 19 x 5 x 19 = 95 x 95, since 95 = 5 x 19

So we get: 9025/95 = 95, and thus, 9025/(-95) = -95.

Note again though, you don't have to do calculations the way above.
And the same is true, too, for all the other examples.
There can be many ways to calculate.

B.   5625 / -75

75 = 70 + 5 = 7 x 2 x 5 + 5 = 14 x 5 + 5 = (14 + 1) x 5 = 15 x 5 = 3 x 5 x 5

5625 = 5000 + 600 + 25
5000 = 5 x 1000 = 5 x 5 x 200 = <u>25 x 200</u>
600 = 6 x 100 = 6 x 25 x 4 = **25 x 24**

So 5625 = <u>25 x 200</u> + **25 x 24** + 25 = 25 x (200 + 24 + 1) = 25 x 225

225 = 200 + 25
200 = 2 x 100 = 2 x 25 x 4 = 25 x 8

So 225 = 25 x 8 + 25 = 25 x (8 + 1) = 25 x 9

5625 = 25 x 225 = 25 x 25 x 9, and 75 = 3 x 5 x 5 = 3 x 25
5625 = 25 x 25 x 3 x 3, 75 = 25 x 3, and 5625 = 75 x 75

So we get: 5625/75 = 75, and thus, 5625/(-75) = -75.

C.   7225 / -85

7225 = 7000 + 225
7000 = 7 x 1000 = 7 x 5 x 200 = 7 x 5 x 2 x 100 = 7 x 5 x 2 x 4 x 25
225 = 200 + 25 = 2 x 4 x 25 + 25 = 8 x 25 + 25 = 9 x 25

7225 = 7 x 5 x 2 x 4 x 25 + 9 x 25 = 7 x 4 x 2 x 5 x 25 + 9 x 25 = 28 x 10 x 25 + 9 x 25
= 280 x 25 + 9 x 25 = 289 x 25

85 = 80 + 5, and 80 = 8 x 10 = 8 x 2 x 5 = 16 x 5
So 85 = 16 x 5 + 5 = 17 x 5

And we have: 289 = 17 x 17
So 7225 = 289 x 25 = 17 x 17 x 25 = 17 x 17 x 5 x 5 = 17 x 5 x 17 x 5 = 85 x 85

Therefore, 7225/85 = 85, and 7225/(-85) = -85.

D.   -4225 / 65

4225 = 4000 + 225

4000 = 4 x 1000 = 4 x 10 x 100 = 4 x 10 x 4 x 25 = 16 x 10 x 25 = 160 x 25

225 = 200 + 25 = 2 x 4 x 25 + 25 = 8 x 25 + 25 = 9 x 25

4225 = 160 x 25 + 9 x 25 = (160 + 9) x 25 = 169 x 25

65 = 60 + 5 = 6 x 2 x 5 + 5 = 12 x 5 + 5 = 13 x 5, and 169 = 13 x 13

4225 = 169 x 25 = 13 x 13 x 25 = 13 x 13 x 5 x 5 = 13 x 5 x 13 x 5 = 65 x 65

So we get: 4225/65 = 65, and –4225/65 = -65

E.   123123 / -1001

123123 = 123000 + 123 = 123 x 1000 + 123 = 123 x (1000 + 1) = 123 x 1001

So we get: 123123/1001 = 123, and 123123/(-1001) = -123.

F.   1230123 / -10001

1230123 = 1230000 + 123 = 123 x 10000 + 123 = 123 x (10000 + 1) = 123 x 10001

So we get: 1230123/10001 = 123, and 1230123/(-10001) = -123

G.   -12300123 / 100001

12300123 = 12300000 + 123 = 123 x 100000 + 123 = 123(100000 + 1) = 123 x 100001

So we get: 12300123/100001 = 123, and -12300123/100001 = -123.

Examples H

Do the divisions below:

0.   -1 / (-1)

1.   -2 / (-2)

2.   -44 / (-11)

3.   -3232 / (-101)

4.   -32032 / (-1001)

5.   -321321 / (-1001)

6.   -121 / (-11)

7.   -12321 / (-111)

8.   -24642 / (-222)

9.   -36963 / (-333)

A.   -1234231 / (-1111)

B.   -2468642 / (-2222)

C.   -36963 / (-111)

Suggestions or Solutions
To the Examples H

In division, if the <u>number of negatives is odd</u>, the quotient is <u>negative</u>.
And if the number of negatives is even, the quotient is positive.　So for instance:

$24 \div (-4) \div (-2) = 3$, that is, $24/(-4)/(-2) = 3$.

$-37/(-7)/5/(-2) = -37/70$, and $-37/(-7)/(-5)/(-2) = 37/70$.

So in general, doing divisions of numbers positive and negative mixed together, we can just take all the numbers as positive numbers, take the quotient, and then, add the negative sign to the quotient if odd is the number of all the negatives.

For instance, doing this: $-17/(-9)/5/(-2)$, do this: $-17/9/5/2$.

And doing this: $-28/(-5)/(-3)/(-2)$, just do this: $28/5/3/2$, because the number of negatives is 4, which is even.

0.　$-1 / (-1) = 1$

1.　$-2 / (-2) = \dfrac{-2}{-2} = 1$

2.　$-44 / (-11) = \dfrac{-44}{-11} = \dfrac{-1 \times 44}{-1 \times 11} = \dfrac{44}{11} = \dfrac{4 \times 11}{11} = \dfrac{4}{1} = 4$.

3.　$-3232 / (-101) = \dfrac{-1 \times 3232}{-1 \times 101} = \dfrac{3232}{101} = \dfrac{32 \times 101}{101} = \dfrac{32}{1} = 32$.

4.   -32032 / (-1001)

32032 = 32000 + 32 = 32 x 1000 + 32 = 32 x 1001

So we get: -32032 / (-1001) = $\dfrac{-32032}{-1001} = \dfrac{32032}{1001} = \dfrac{32 \times 1001}{1001} = \dfrac{32}{1} = 32$.

And we can write it this way, too: -32032/(-1001) = 32032/1001 = 32 x 1001/1001 = 32.

5.   -321321 / (-1001)

321321 = 321000 + 321 = 321 x 1000 + 321 = 321 x 1001

So we get: -321321 / (-1001) = $\dfrac{-1 \times 32132}{-1 \times 1001} = \dfrac{32132}{1001} = \dfrac{321 \times 1001}{1001} = \dfrac{321}{1} = 321$.

It can be written this way, too: -321321/(-1001) = 321321/1001 = 321·1001/1001 = 321.

6.   -121 / (-11)

121 = 100 + 20 + 1 = 100 + 10 + 10 + 1 = 10 x 10 + 10 + 10 + 1

And we have: 10 x 10 + 10 = 10 x (10 + 1) = 10 x 11

So we get: 121 = 10 x 11 + 10 + 1 = 10 x 11 + 11 = (10 + 1) x 11 = 11 x 11

So we get: -121 / (-11) = $\dfrac{-121}{-11} = \dfrac{121}{11} = \dfrac{11 \times 11}{11} = \dfrac{11}{1} = 11$.

And of course, it can be put this way, too: -121/(-11) = 121/11 = 11 x 11/11 = 11.

Note however, you don't have to do calculations the way above.

7. -12321 / (-111)

$12321 = 10000 + 2000 + 300 + 20 + 1$

$= 10000 + 1000 + 1000 + 100 + 100 + 100 + 10 + 10 + 1$

$= 10000 + 1000 + 100 + 1000 + 100 + 10 + 100 + 10 + 1$

$10000 + 1000 + 100 = 100 \times (100 + 10 + 1)$

$1000 + 100 + 10 = 10 \times (100 + 10 + 1)$

$12321 = 100 \times (100 + 10 + 1) + 10 \times (100 + 10 + 1) + 100 + 10 + 1$
$= 100 \times 111 + 10 \times 111 + 111$
$= 111 \times (100 + 10 + 1)$
$= 111 \times 111$

So we get: $-12321 / (-111) = \dfrac{-12321}{-111} = \dfrac{12321}{111} = \dfrac{111 \times 111}{111} = \dfrac{111}{1} = 111$.

And we can write it this way, too: $-12321/(-111) = 12321/111 = 111 \times 111/111 = 111$.

8. -24642 / (-222)

$24642 = 2 \times 12321$.   And $12321 = 111 \times 111$ {as shown in the 7 above}

$24642 = 2 \times 111 \times 111 = 222 \times 111$

So we get: $-24642 / (-222) = \dfrac{-24642}{-222} = \dfrac{24642}{222} = \dfrac{222 \times 111}{222} = \dfrac{111}{1} = 111$.

And we can put it this way, too: $-24642/(-222) = 24642/222 = (222 \times 111)/222 = 111$.

9.   -36963 / (-333)

36963 = 3 x 12321 = 3 x 111 x 111   {as shown in the 7 above}

So 36963 = 333 x 111.

Thus, we get: -36963 / (-333) = $\dfrac{-36963}{-333} = \dfrac{36963}{333} = \dfrac{333 \times 111}{333} = \dfrac{111}{1} = 111$.

And we can put it this way, too: -36963/(-333) = 36963/333 = (333 x 111)/333 = 111.

A.   -1234231 / (-1111)

1,234,321 = 1,000,000 + 200,000 + 30,000 + 4000 + 300 + 20 + 1

= 1,000,000 + 100000 + 10000 + 1000 +
                              100000 + 10000 + 1000 + 100 +
                                          10000 + 1000 + 100 + 10 +
                                                      1000 + 100 + 10 + 1

= 1000 x (1000 + 100 + 10 + 1) +
      100 x (1000 + 100 + 10 + 1) +
        10 x (1000 + 100 + 10 + 1) +
            1000 + 100 + 10 + 1

= 1000 x 1111 + 100 x 1111 + 10 x 1111 + 1111
= 1111 x (1000 + 100 + 10 + 1)
= 1111 x 1111

So we get: -1234321 / (-1111) = $\dfrac{-1234321}{-1111} = \dfrac{1234321}{1111} = \dfrac{1111 \times 1111}{1111} = \dfrac{1111}{1} = 1111$.

And of course, we can put it this way, too:

-1234321/(-1111) = 1234321/1111 = (1111 x 1111)/1111 = 1111.

98

B. -2468642 / (-2222)

2468642 = 2 x 1234321 = 2 x 1111 x 1111   {as shown in the example A above}

So we get: 2468642 = 2222 x 1111.

So we get: -2468642 / (-2222) = $\dfrac{-2468642}{-2222} = \dfrac{2468642}{2222} = \dfrac{2222 \times 1111}{2222} = \dfrac{1111}{1} = 1111$.

And of course, we can put it this way, too:
-2468642/(-2222) = 2468642/2222 = (2222 x 1111)/2222 = 1111.

C. -36963 / (-111)

36963 = 3 x 12321 = 3 x 111 x 111   {as shown in the example 7 above}

So 36963 = 333 x 111.

So we get: -36963 / (-111) = $\dfrac{-36963}{-111} = \dfrac{36963}{111} = \dfrac{333 \times 111}{111} = \dfrac{333}{1} = 333$

And of course, we can put it this way, too:
-36963/(-111) = 36963/111 = (333 x 111)/111 = 333.

Examples I

Do the divisions below:

0. -44 ÷ (-11) ÷ 4 = -44 / (-11) / 4

1. -3232 / (-32) / (-101)   2. -32032 / (-1001) / (-1)

3. -321321 / 321 / (-1001)   4. -121 / (-11) / (-11) / (-1)

5. -12321 / (-111) / (-123) / (-1) 6. - 24642 / (-111) / (-2)

7. -36963 / (-111) / (-3)   8. -1234231 / (-1111) / (-101)

9. -2468642 / (-1111) / (-2) / (-101)   A. -2468642 / (-1111) / (-101)

B. -36963 / (-111) / (-3)   C. 121 / (-11) / (-1)

D. 225 / (-25) / (-5)   E. -625 / 25 / (-5)

F. -1225 / 35 / (-5)   G. -2025 / 45 / (-5)

H. 3025 / -55 / (-11)   I. 9025 / -95 / (-5)

J. 5625 / -75 / (-15)   K. 7225 / -85 / (-5)

L. -4225 / 65 / (-13)   M. 123123 / -1001 / (-123)

N. 1230123 / -10001 / (-123)   O. -12300123 / 100001 / (-123)

Suggestions or Solutions
To the Examples I

In division, if the <u>number of negatives is odd</u>, the quotient is <u>negative</u>.
And if the number of negatives is even, the quotient is positive.   So for instance:

$24 \div (-4) \div (-2) = 3$, that is, $24/(-4)/(-2) = 3$, and $-37/(-7)/5/(-2) = -37/70$.

So in general, doing divisions of numbers positive and negative mixed together, we can just take all the numbers as positive numbers, take the quotient, and then, add the negative sign to the quotient if odd is the number of all the negatives.

So for instance, doing this: $-17/(-9)/5/(-2)$, do this: $-17/9/5/2$.
And doing this: $-28/(-5)/(-3)/(-2)$, just do this: $28/5/3/2$.

0. $-44 / (-11) / 4 = -44 \div (-11) \div 4$.

In this case, we do **$-44 \div (-11)$** first, and then, divide the result, that is, the quotient by 4.

And in general, doing a division as $A/B/C/D\ldots$, we do division operation from the left to the right in such a sequential manner as described above.

So assuming $U = A/B$, $V = U/C$, and $W = V/D$, we get: $W = A/B/C/D$.

In other words, getting W, that is, doing $A/B/C/D$, we do first: A/B to get U, then do U/C to get V, and then do V/D to get W.   And we can put W the way below, too:

$W = A/(BCD) = \dfrac{A}{BCD}$. And doing: $A/B/C/D/E\ldots$, we can do: $\dfrac{A}{BCDE\ldots}$.

So to begin with, we get first: $-44/(-11) = 44/11 = 4$, and then get: $4/4 = 1$.
In sum, we get: $-44 / (-11) / 4 = 4/4 = 1$.

And using the fact above, we can get it the way below, too:
$-44 / (-11) / 4 = -44 / (-11 \times 4) = -44/(-44) = 1$.

1.   -3232 / (-32) / (-101)

To begin with, we have: $3232 = 3200 + 32 = 32 \times 100 + 32 = 32 \times (100 + 1) = 32 \times 101$.
So first, we get: $-3232 / (-32) = 3232/32 = 101$.
And next, we get: $101/(-101) = -1$.

In sum, we get: $-3232 / (-32) / (-101) = 101/(-101) = -1$.

And using the fact stated in the example 0, we can get it this way, too:
$-3232 / (-32) / (-101) = -3232 / \{-32 \times (-101)\} = -3232 / 3232 = -1$.

And we know the fact that in division, if the <u>number of negatives is odd</u>, the quotient is <u>negative</u>, and if the number of negatives is even, the quotient is positive.

So using the fact above and the fact stated in the example 0, we can get it this way, too:
$-3232 / (-32) / (-101) = -3232/32/101 = -3232/(32 \times 101) = -3232 / 3232 = -1$.

2.   -32032 / (-1001) / (-1)

To begin with, $32032 = 32000 + 32 = 32 \times 1000 + 32 = 32 \times (1000 + 1) = 32 \times 1001$.
So we can get first: $-32032 / (-1001) = 32032/1001 = 32$.
And next, we get: $32/(-1) = -32$.

In sum, we get: $-32032 / (-1001) / (-1) = 32/(-1) = -32$.

And we can get it the way below, too:

$-32032 / (-1001) / (-1) = -32032/\{-1001 \times (-1)\} = -32032/1001 = (-1 \times 32032)/1001$
$= (-1 \times 32 \times 1001)/1001 = -1 \times 32 \times (1001/1001) = -1 \times 32 \times 1 = -1 \times 32 = -32$.

And using the facts stated above, we can get it the way below, too:

$-32032 / (-1001) / (-1) = -32032/1001/1 = -32032/(1001 \times 1) = -32032/1001$
$= -32 \times 1001/1001 = -32$.

3.   -321321 / 321 / (-1001)

To begin with, we can get: 321321 = 321 x 1000 + 321 = 321 x 1001.
So we get first: -321321/321 = -1001.   And next, we get: -1001/(-1001) = 1.

In sum, we get: -321321 / 321 / (-1001) = -1001 / (-1001) = 1.

And using the facts stated above, we can get it the way below, too:

-321321 / 321 / (-1001) = 321321/321/1001 = 321321/(321 x 1001)
= 321321/321321 = 1.

4.   -121 / (-11) / (-11) / (-1)

To begin with, we can get:

121 = 100 + 10 + 10 + 1 = 10 x 10 + 10 + 10 + 1 = 10 x (10 + 1) + 10 + 1
= 10 x 11 + 10 + 1 = 10 x 11 + 11 = 11 x 11

So first, we can get: -121 / (-11) = 121/11 = (11 x 11)/11 = 11.
Thus next, we can get: 11 / (-11) = -1.   And next, we get: -1 / (-1) = 1.

And using the facts stated above, we can get it this way, too:

-121 / (-11) / (-11) / (-1) = 121/11/11/1 = 121/(11 x 11 x 1) = 121/121 = 1.

5.   -12321 / (-111) / (-123) / (-1)
= 12321/111/123/1 = 12321/(111 x 123 x 1) = 12321/12321 = 1.

6.   -24642 / (-111) / (-2) = -24642/111/2 = -24642/(111 x 2) = -24642/222.

And we can get: 24642 = 2 x 12321 = 2 x 111 x 111 = 222 x 111.

So we get: -24642/222 = -111, which equals, of course, -24642 / (-111) / (-2).

7. -36963 / (-111) / (-3) = -36963/111/3 = -36963/(111 x 3) = -36963/333.

And we can get: 36963 = 3 x 12321 = 3 x 111 x 111 = 333 x 111.
So we get: -36963/333 = -111.

8. -1234321 / (-1111) / (-101) = -1234321/1111/101.

And we can get: 1234321 = 1111 x 1111.   So we get: 1234321/1111 = 1111.

Next, 1111 = 1100 + 11 = 11 x 100 + 11 = 11 x 101.   So we get: 1111/101 = 11.

Thus, we get: -1234321 / (-1111) / (-101) = -1234321/1111/101 = 11.

And we can get the same the way below, too:

1111 x 101 = 1111 x (100 + 1) = 111100 + 1111 = 112211

1234321 − 112211 = 1000000 + 100000 + 20000 + 2000 + 100 + 10 = 1122110
= 112211 x 10.   That is to say that 1234321 − 112211 = 112211 x 10.

So we get: 1234321 = 112211 x 10 + 112211 = 112211 x (10 + 1) = 112211 x 11.

And thus, we get:

-1234321 / (-1111) / (-101) = -1234321/1111/101 = -1234321/(1111 x 101)
= -1234321 / 112211 = -(112211 x 11)/112211 = -11.

9. -2468642 / (-1111) / (-2) / (-101) = 2468642/1111/2/101.

This time, let's do it the way below:

First, we do this: 2468642/1111.   Next, using the result, we do: 2468642/1111/2.

And then, using the result, we do: 2468642/1111/2/101.

We can get first: $2468642 = 2 \times 1234321 = 2 \times 1111 \times 1111 = 2222 \times 1111$.

So we get: $2468642/1111 = 2222$.   Next, $2222/2 = 1111$.

And we can get: $1111 = 1100 + 11 = 11 \times 100 + 11 = 11 \times 101$.

So next, we get: $1111/101 = 11 \times 101/101 = 11$.

In sum, we get:

$-2468642 / (-1111) / (-2) / (-101) = \underline{2468642/1111}/2/101 = \underline{2222}/2/101 = 1111/101 = 11$.

A.   -2468642 / (-1111) / (-101) $= -2468642/1111/101$.

This time, too, let's do it the way above.

To begin with, $2468642 = 2 \times 1234321 = 2 \times 1111 \times 1111 = 2222 \times 1111$.

So we get: $2468642/1111 = 2222 \times 1111/1111 = 2222$.

And $2222 = 2 \times 1111 = 2 \times 11 \times 101 = 22 \times 101$. So next, $2222/101 = 22 \times 101/101 = 22$.

In sum, we get:

$-2468642 / (-1111) / (-101) = -2468642/1111/101 = -2222/101 = -22$.

B.   -36963 / (-111) / (-3) $= -36963/111/3$.

This time, too, let's do it the way above.

To begin with, we can get: $36963 = 3 \times 12321 = 3 \times 111 \times 111 = 333 \times 111$.

So first, we get: $36963/111 = 333$. And next, we get: $333/3 = 111$.

So in sum, we get: $-36963 / (-111) / (-3) = -36963/111/3 = -333/3 = -111$.

C.   121 / (-11) / (-1) = 121/11/1 = 121/11 = 11.

D.   225 / (-25) / (-5) = 225/25/5 = 225/(25 x 5) = 25 x 9/(25 x 5) = 9/5.

E.   -625 / 25 / (-5) = 625/25/5.

First, 625 = 600 + 25 = 6 x 100 + 25 = 6 x 4 x 25 + 25 = 24 x 25 + 25 = 25 x 25.
So we get: 625/25 = 25.   Thus we get: 625/25/5 = 25/5 = 5.

And we can get it the way below, too:

We can get: 625/25/5 = 625/(25 x 5).   And we have: 625 = 25 x 25
So next, we get: 625/(25 x 5) = 25 x 25/(25 x 5) = 5.

F.   -1225 / 35 / (-5) = 1225/35/5.

First, we can get:
$1225 = 1200 + 25 = 12 \times 100 + 25 = 12 \times 4 \times 25 + 25 = 48 \times 25 + 25 = 25 \times 49 = 5^2 \times 7^2$.

And we have: 35 = 5 x 7, so we can get: $1225 / 35 = 5^2 \times 7^2 / (5 \times 7) = 5 \times 7$.
Thus, we get: 1225/35/5 = 5 x 7/5 = 7.

And we can notice that $1225 = 35^2$, because $1225 = 5^2 \times 7^2 = (5 \times 7)^2 = 35^2$.
So we can get it the way below, too: 1225/35/5 = 35/5 = 7.

G.   -2025 / 45 / (-5) = 2025/45/5.

$2025 = 2000 + 25 = 20 \times 100 + 25 = 20 \times 4 \times 25 + 25 = 25 \times 80 + 25 = 25 \times 81 = 5^2 9^2$.

And $5^2 9^2 = 45^2$.   So we get: 2025/45/5 = 45/5 = 9.

H. 3025 / -55 / (-11) = 3025/55/11.

First, $3025 = 3000 + 25 = 30 \times 100 + 25 = 30 \times 4 \times 25 + 25 = 25 \times 120 + 25 = 25 \times 121$.
And $25 \times 121 = 25 \times 11 \times 11 = 5^2 \times 11^2 = 55^2$.

So we get: 3025/55/11 = 55/11 = 5.

I. 9025 / -95 / (-5) = 9025/95/5.

First, $9025 = 9000 + 25 = 90 \times 100 + 25 = 90 \times 4 \times 25 + 25 = 25 \times 360 + 25 = 25 \times 361$.
And $361 = 19 \times 19$.
So we get: $9025 = 25 \times 19 \times 19 = 5^2 \times 19^2 = (5 \times 19)^2 = 95^2$.
Thus, we get: 9025/95/5 = 95/5 = 19.

J. 5625 / -75 / (-15) = 5625/75/15.

First, $5625 = 5000 + 625 = 50 \times 100 + 25 \times 25 = 25 \times 200 + 25 \times 25 = 25 \times 225$.
And $225 = 200 + 25 = 8 \times 25 + 25 = 25 \times 9$.
So $5625 = 25 \times 225 = 25 \times 25 \times 9$.

Next, $75 = 25 \times 5$.

So we get: $5625/75/15 = 25 \times 25 \times 9/(25 \times 5)/15 = 25 \times 9/5/15 = 5 \times 9/15 = 9/3 = 3$.

K. 7225 / -85 / (-5) = 7225/85/5.

First, $85 = 80 + 5 = 16 \times 5 + 5 = 5 \times 17$.
Next, $7225 = 7000 + 225 = 70 \times 4 \times 25 + 25 \times 9 = 25 \times 280 + 25 \times 9 = 25 \times 289$.
And $289 = 17 \times 17$. So we get: $7225 = 25 \times 17 \times 17$.
So we get: $7225/85/5 = 25 \times 17 \times 17/(5 \times 17)/5 = 25 \times 17 \times 17/(25 \times 17) = 17$.

L.   **-4225 / 65 / (-13)** = 4225/65/13.

First, 65 = 60 + 5 = 12 x 5 + 5 = 5 x 13.
Next, 4225 = 4000 + 225 = 40 x 100 + 25 x 9 = 160 x 25 + 25 x 9 = 25 x 169.
And 169 = 13 x 13.   So we get: 4225 = 25 x 13 x 13.
So we get: 4225/65/13 = 25 x 13 x 13/(5 x 13)/13 = 25 x 13 x 13/(5 x 13 x 13) = 5.

M.   **123123 / -1001 / (-123)** = 123123/1001/123.

First, 123123 = 123 x 1001.
So we get: 123123/1001/123 = 123 x 1001/1001/123 = 123/123 = 1.

N.   **1230123 / -10001 / (-123)** = 1230123/10001/123.

First, 1230123 = 1230000 + 123 = 123 x 10000 + 123 = 123 x 10001.
So we get: 1230123/10001/123 = 123/123 = 1.

O.   **-12300123 / 100001 / (-123)** = 12300123/100001/123.

First, 12300123 = 12300000 + 123 = 123 x 100000 + 123 = 123 x 100001.
So we get: 12300123/100001/123 = 123/123 = 1.

Examples J

0.   Find the number of all positive integers that are less than 784, and have 7 in the 100's digit, and 3 in the 1's digit.

1.   Find the number of all positive integers that are less than 1004, and have 8 in the 100's digit, and 5 in the 1's digit.

2.   Put the numbers in the list below in ascending order, that is, put the smallest number first and the largest number last: 7852, 7694, 7926, 7799, 7901, and 7890.

3.   Assuming *d* is a single digit number, find *d* that makes 59*d*4 greater than 5*d*99.

4.   Assuming *d* is a single digit number, find *d* that makes 8*d*09 greater than 89*d*5.

5.0.   Find all positive integers that can satisfy all below at the same time:

   A. The 100's digit is 3 larger than the 10's digit.
   B. The 10's digit is 1 lager than the 1's digit.
   C. The 1's digit is lager than 2.
   D. The integers are between 3000 and 4000.

5.1.   Find all positive integers that can satisfy all below at the same time:

   A. The 100's digit is 1 larger than the 10's digit.
   B. The 10's digit is 2 larger than the 1's digit.
   C. The 1's digit is lager than 1.
   D. The integers are between 3000 and 5000.

5.2.   Find all positive integers that can satisfy all below at the same time:

   A. The 100's digit is 3 smaller than the 10's digit.
   B. The 10's digit is 2 larger than the 1's digit.
   C. The 1's digit is smaller than 7.
   D. The integers are between 3000 and 5000.

Suggestions or Solutions
To the Problem in the Example 0

Find the number of all positive integers that are less than 784, and have 7 in the 100's digit, and 3 in the 1's digit.

There are 9 integers.

If not sure of how to get it, follow the steps below:

Assuming first, A is a positive integer less than 784, we can say that A can be one of all the integers from 1 to 783.

So assuming next, B is a positive integer that is less than 784, and has 7 in the 100's digit, and 3 in the 1's digit, we can set: $B = 7D3$ where D is the 10's digit.

What number then, can be D?

As each digit in a number, we can use one of ten single digit numbers, which are 0, 1, 2, 3, …, 9.

And we know that B is an integer positive and less than 784.
And we set: $B = 7D3$ where D is the number in the 10's digit.

So we can use as D one of 9 numbers that are 0, 1, 2, 3, …, 8.

How many integers then, can we use as B?

We can use 9 numbers as D. And thus, there are 9 integers that can be B.

They are 783, 773, 763, 753, 743, 733, 723, 713, and 703.

Suggestions or Solutions
To the Problem in the Example 1

Find the number of all positive integers that are less than 1004, and have 8 in the 100's digit, and 5 in the 1's digit.

There are 10 integers.

If not sure of how to get it, follow the steps below:

Assuming first, A is a positive integer less than 1004, we can say that A can be one of all the integers from 1 to 1004.

So assuming next, B is a positive integer that is less than 1004, and has 8 in the 100's digit, and 5 in the 1's digit, we can set: $B = T8D5$ where T is the 1000's digit, and D is the 10's digit.

Then first, what can be T?

As each digit in a number, we can use one of ten single digit numbers, which are 0, 1, 2, 3, ..., 9.

And we know that B is an integer positive and less than 1004.
So we can use 0 only as T.

And thus, we can now set: $B = 8D5$.

And we can use as D one of ten numbers, which are these: 0, 1, 2, 3, ..., 0.

So there are 10 integers that can be B.

And they are 895, 885, 875 . . . 815, and 805.

Suggestions or Solutions
To the Problem in the Example 2

Put the numbers in the list below in ascending order, that is, put the smallest number first and the largest number last: 7852, 7694, 7926, 7799, 7901, and 7890.

So putting all those numbers in ascending order, we can put them the way below:

7694, 7799, 7852, 7901, and **7926**.

It won't probably take too long to get it, because we can look at all the numbers in the list, and the list is not too long.   *If want to see how though, follow the steps below:*

If positive, the smaller the digit, the smaller the number.

So what number can be the smallest in a list of positive numbers if all the numbers have the same number of digits?

Of the numbers positive, the smallest can be the number where the highest digit is the smallest if all the numbers have the same number of digits.

So in short, if the highest digit is the smallest, the number can be the smallest.

What if however, all the highest digits in all the numbers are the same?

Then, we check the next highest digit, that is, the second highest digit, which is in this case, the hundred's digit.

And if positive, the smaller the digit, the smaller the number.
So what is the smallest of all in the list below?

7852, 7694, 7926, 7799, 7901, and **7890**.

We can see that **7694** has 6 in the 100's digit, and that 6 is the smallest of all the 100's digit.

So **7694** is the smallest in the list above.

And the second smallest is **7799**.

And thus, eliminating the two from the list, we now have: **7852, 7926, 7901,** and **7890**.

Next, we can see that the two numbers 7852 and 7890 have 8 in the 100's digit.
So in the two, we need to look at the next lower digit, which is the 10's digit.

Then, 5 is smaller than 9, so **7852** is the smaller of the two, and thus, is the third smallest.

Then, of course, **7890** is the fourth smallest.

And thus, we are now left with **7926** and **7901**, of which 7901 is the smaller.

So putting all the numbers in the list in ascending order, we can put them the way below:

7694, 7799, 7852, 7901, and **7926.**

Suggestions or Solutions
To the Problem in the Example 3

Assuming *d* is a single digit number, find *d* that makes 59*d*4 greater than 5*d*99.

We get: *d* = 0, 1, 2 . . . 7, or 8.

If not sure of how to get it, follow the steps below:

We want 59*d*4 to be larger than 5*d*99.

And if positive, the larger the digit, the larger the number.

So we may want to begin with checking the highest digits in the two numbers given.

To begin with, the 1000's digits are the same.

So moving on to the next lower digit, which is the 100's digit, we have 9 and *d*.

And we want 59*d*4 to be larger than 5*d*99.   So for now, what does *d* have to be?

We can see for now, that *d* has to be less than or equal to 9.
Thus, we get: *d* = 0, 1, 2 . . . 8, or 9.

And of course, we want to check to see if all the numbers can work for *d*.

If *d* is one of 0, 1, 2 . . . 7, and 8, we get: 59*d*4 > 5*d*99.
If however, *d* = 9, we cannot get: 59*d*4 > 5*d*99.   So we get: *d* ≠ 9.

And thus, we get: *d* = 0, 1, 2 . . . 7, or 8.

Suggestions or Solutions
To the Problem in the Example 4

Assuming *d* is a single digit number, find *d* that makes 8*d*09 greater than 89*d*5.

There is no value for *d*.

If want to see why not, follow the steps below:

We want 8*d*09 to be larger than 89*d*5.

And if positive, the larger the digit, the larger the number.
So we may want to begin with checking the highest digits in the two numbers given.

To begin with, the 1000's digits are the same.
So moving on to the next lower digit, which is the 100's digit, we have *d* and 9.

And we want 8*d*09 to be larger than 89*d*5. So for now, what *d* has to be?

We can see for now, that *d* has to be equal to 9, since *d* is a digit, which is one of 0, 1, 2, …, 8, or 9.

And thus, assuming *d* = 9, we get: 8*d*09 = 8909, and 89*d*5 = 8995, which is however, not the case we want, because we want this: 8*d*09 > 89*d*5 for some *d*.

So there is no value for *d*.

Suggestions or Solutions
To the Problem 0 in the Example 5

Find all positive integers that can satisfy all below at the same time:

    A. The 100's digit is 3 larger than the 10's digit.
    B. The 10's digit is 1 lager than the 1's digit.
    C. The 1's digit is lager than 2.
    D. The integers are between 3000 and 4000.

   The numbers are as follows: **3743**, **3854**, and **3965**.

If not sure of how to get it, follow the steps below:

Assuming first, that n is an integer described above, we can say from the fact D, that n is a 4-digit integer.
So we can set: $n = thdu$, where t is the 1000's digit, h is the 100's digit, d is the 10's digit, and u is the 1's digit. Then first, what does t have to be?

We can say that $t = 3$, since n is between 3000 and 4000.

Now, we have set: $n = thdu$, where t is the 1000's digit, h is the 100's digit, d is the 10's digit, and u is the 1's digit. And we have the facts below:

    A. The 100's digit is 3 larger than the 10's digit.
    B. The 10's digit is 1 lager than the 1's digit.
    C. The 1's digit is lager than 2.

So next, from the fact **A**, what can we say about h and d?

The fact **A** is saying that the 100's digit is 3 larger than the 10's digit.
And we have set: $n = thdu$, where t is the 1000's digit, h is the 100's digit, d is the 10's digit, and u is the 1's digit. So from the fact **A**, we can say that $h = 3 + d$.

And we have the other two facts below:

B.   The 10's digit is 1 lager than the 1's digit.
C.   The 1's digit is lager than 2.

So next, from the fact **B**, what can we say about d and u?

The fact **B** is saying that the 10's digit is 1 lager than the 1's digit.

And we have set: $n = thdu$, where t is the 1000's digit, h is the 100's digit, d is the 10's digit, and u is the 1's digit.   So from the fact **B**, we can say that $d = 1 + u$.

And next, we have the fact **C**, which is saying that the 1's digit is lager than 2.
So next, from the fact **C**, what can we say about u?

The fact **C** is saying that the 1's digit is lager than 2.

And we set: $n = thdu$, where t is the 1000's digit, h is the 100's digit, d is the 10's digit, and u is the 1's digit.
So from the fact **C**, we can say that $u > 2$. Thus for now, we get: $u = 3, 4, 5, 6, 7, 8,$ or 9.

So to begin with, we get: $u = 3 \Rightarrow d = 1 + u = 4 \Rightarrow h = 3 + d = 7$.

Next, we get: $u = 4 \Rightarrow d = 1 + u = 5 \Rightarrow h = 3 + d = 8$.

Next, we get: $u = 5 \Rightarrow d = 1 + u = 6 \Rightarrow h = 3 + d = 9$.

And next, we get: $u = 6 \Rightarrow d = 1 + u = 7 \Rightarrow h = 3 + d = 10$, which is however, is not allowed, since h is a digit.

So we get: $(u, d, h) = (3, 4, 7), (4, 5, 8),$ and $(5, 6, 9)$.

And we know: $n = thdu$, where $t = 3$.   So we get: $n = $ **3743**, **3854**, or **3965**.

Let's now, put all the ideas together in one page.

5.0. **Find all positive integers that can satisfy all below at the same time:**

    A. **The 100's digit is 3 larger than the 10's digit.**
    B. **The 10's digit is 1 lager than the 1's digit.**
    C. **The 1's digit is lager than 2.**
    D. **The integers are between 3000 and 4000.**

Assuming first, that n is an integer described above, we can say from the fact **D**, that n is a 4-digit integer.

So we can set: $n = thdu$, where t is the 1000's digit, h is the 100's digit, d is the 10's digit, and u is the 1's digit.

And also, we can see that $t = 3$, since n is between 3000 and 4000.

Next, from the fact **A**, we can say that $h = 3 + d$.

Next, from the fact **B**, we can say that $d = 1 + u$.

And next, from the fact **C**, we can say that $u > 2$.
Thus for now, we get: $u = 3, 4, 5, 6, 7, 8,$ or 9.

So to begin with, we get: $u = 3 \Rightarrow d = 1 + u = 4 \Rightarrow h = 3 + d = 7$.

Next, we get: $u = 4 \Rightarrow d = 1 + u = 5 \Rightarrow h = 3 + d = 8$.

Next, we get: $u = 5 \Rightarrow d = 1 + u = 6 \Rightarrow h = 3 + d = 9$.

And next, we get: $u = 6 \Rightarrow d = 1 + u = 7 \Rightarrow h = 3 + d = 10$, which is however, is not allowed since h is a digit.

So we get: $(u, d, h) = (3, 4, 7), (4, 5, 8),$ and $(5, 6, 9)$.

And we know: $n = thdu$, where $t = 3$.

So we get: $n = $ **3743, 3854,** or **3965**.

Suggestions or Solutions
To the Problem 1 in the Example 5

Find all positive integers that can satisfy all below at the same time:

    A. The 100's digit is 1 larger than the 10's digit.
    B. The 10's digit is 2 larger than the 1's digit.
    C. The 1's digit is lager than 1.
    D. The integers are between 3000 and 5000.

Assuming first, that n is an integer described above, we can say from the fact **D**, that n is a 4-digit integer.
So we can set: $n = thdu$, where t is the 1000's digit, h is the 100's digit, d is the 10's digit, and u is the 1's digit.
And also, we can see that $t = 3$ or 4, since n is between 3000 and 5000.

Next, from the fact **A**, we can say that $h = 1 + d$.
Next, from the fact **B**, we can say that $d = 2 + u$.
And next, from the fact **C**, we can say that $u > 1$. So we get: $u = 2, 3, 4, 5, 6, 7, 8,$ or 9.

So to begin with, we get: $u = 2 \Rightarrow d = 2 + u = 4 \Rightarrow h = 1 + d = 5$.

Next, we get: $u = 3 \Rightarrow d = 2 + u = 5 \Rightarrow h = 1 + d = 6$.

Next, we get: $u = 4 \Rightarrow d = 2 + u = 6 \Rightarrow h = 1 + d = 7$.

Next, we get: $u = 5 \Rightarrow d = 2 + u = 7 \Rightarrow h = 1 + d = 8$.

Next, we get: $u = 6 \Rightarrow d = 2 + u = 8 \Rightarrow h = 1 + d = 9$.

And next, we get: $u = 7 \Rightarrow d = 2 + u = 9 \Rightarrow h = 1 + d = 10$, which is however, is not allowed, since h is a digit.

So we get: $(u, d, h) = (2, 4, 5), (3, 5, 6), (4, 6, 7), (5, 7, 8),$ and $(6, 8, 9)$.

And we know: $n = thdu$, where $t = 3$ or 4.

So we get: $n = 3542, 4542, 3653, 4653, 3764, 4764, 3875, 4875, 3986,$ or 4986.

Suggestions or Solutions
To the Problem 2 in the Example 5

Find all positive integers that can satisfy all below at the same time:

   A. The 100's digit is 3 smaller than the 10's digit.
   B. The 10's digit is 2 larger than the 1's digit.
   C. The 1's digit is smaller than 7.
   D. The integers are between 3000 and 5000.

Assuming first, that n is an integer described above, we can say from the fact **D**, that n is a 4-digit integer. So we can set: $n = thdu$, where t is the 1000's digit, h is the 100's digit, d is the 10's digit, and u is the 1's digit.

And also, we can see that $t = 3$ or **4**, since n is between 3000 and 5000.

Next, from the fact **A**, we can say that $h = d - 3$.
Next, from the fact **B**, we can say that $d = 2 + u$.
And next, from the fact **C**, we can say that $u < 7$.   So we get: $u = 0, 1, 2, 3, 4, 5,$ or 6.

So to begin with, we get: $u = 0 \Rightarrow d = 2 + u = 2 \Rightarrow h = d - 3 = -1$, which is however, is not allowed since h is a digit.

Next, we get: $u = 1 \Rightarrow d = 2 + u = 3 \Rightarrow h = d - 3 = 0$.

Next, we get: $u = 2 \Rightarrow d = 2 + u = 4 \Rightarrow h = d - 3 = 1$.

Next, we get: $u = 3 \Rightarrow d = 2 + u = 5 \Rightarrow h = d - 3 = 2$.

Next, we get: $u = 4 \Rightarrow d = 2 + u = 6 \Rightarrow h = d - 3 = 3$.

Next, we get: $u = 5 \Rightarrow d = 2 + u = 7 \Rightarrow h = d - 3 = 4$.

And next, we get: $u = 6 \Rightarrow d = 2 + u = 8 \Rightarrow h = d - 3 = 5$.

So we get: $(u, d, h) = (1, 3, 0), (2, 4, 1), (3, 5, 2), (4, 6, 3), (5, 7, 4),$ and $(6, 8, 5)$.

And we know: $n = thdu$, where $t = 3$ or **4**.

So we get: $n = 3031, 4031, 3142, 4142, 3253, 4253, 3364, 4364, 3475, 4475, 3586,$ or 4586.

Examples K

0. What is the largest number?

1. What is the largest positive number?

2. What is the largest negative number?

3. What is the smallest number?

4. What is the smallest positive number?

5. What is the smallest negative number?

6. What is the largest number where there are ten digits, and two of the digits are below (to the right of) the point?

7. What is the largest positive number where there are ten digits, and seven of the digits are above (to the left of) the point?

8. What is the largest negative number where there are ten digits, and four of the digits are below the point?

9. What is the smallest number where there are twelve digits, and three of those are below the point?

A.   What is the smallest positive number that has five digits above the point and seven digits below the point?

B.   What is the smallest negative number that has 10 digits above the point and two digits below the point?

C.   What is the largest 10-digit number where all digits are different, and three of the digits are above the point?

D.   What is the largest 10-digit negative number where all digits are different, and five of the digits are below the point?

E.   What is the smallest 12-digit number where five digits are above the point, and every digit above the point is different from any digit below the point?

F.   Find the smallest 12-digit positive number where 7 digits are above the point, and every digit above the point is different from any of the digits below the point.

G.   Find the smallest 12-digit number where six digits are above the point, and are different, and every digit above the point is different from any digit below the point.

Suggestions or Solutions
To the Problems in the Examples K

0. What is the largest number?

There is no largest number.   Why not, though?

A number can be infinitely large.
So no matter how large number we may come up with, we can still make one the larger.

-9, < -2 < 0 < 1 < 3 < 9 < 9999 < 99999 < 9999999 < 999999999, ...

And thus, there is no largest number.

1. What is the largest positive number?

There is no largest positive number.   Why not?

A positive number can be infinitely large. So no matter how large positive number we may come up with, we can still make the larger one.

0 < 0.0001 < 0.01 < 1 < 3 < 9 < 99 < 9999 < 9999999 < 9999999999999 < … … …

Therefore, there is no largest positive number.

2. What is the largest negative number?

There is no largest number negative.   Why not, though?

A number is negative if it is less than 0.

And the smaller the magnitude, the larger the number if the number is negative.

So a negative number can be indefinitely large, since it is less than 0.   How come?

The magnitude of a negative number can be infinitely small. And the smaller the magnitude, the larger the number if the number is negative.

-9 < -2 < -1 < -0.1 < -0.001 < -0.000001 < -0.0000000000001 < … … … < 0.

And we know 0 is not negative, of course.

So no matter how large negative number we may come up with, we can still make one the larger, which is still negative though. And thus, there is no largest number negative.

3.   What is the smallest number?

   There is no smallest number.   Why not?

A negative number can be infinitely small. And the larger the magnitude, the smaller the number if the number is negative.

99 > 9 > 5 > 1 > 0 > -0.0001 > -0.01 > -1 > -3 > -99 > -999 > -999999999 > … … …

So no matter how small negative number we may come up with, we can still make one the smaller. And thus, there is no smallest number.

4.   What is the smallest positive number?

   There is no smallest positive number.   Why not, though?

A number is positive if it is greater than 0.

And the smaller the magnitude, the smaller the number if the number is positive.

And the magnitude of the positive number can be infinitely small.

So a positive number can be infinitely small although it is greater than 0.

$$9 > 5 > 1 > 0.1 > 0.0001 > 0.0000001 > 0.000000000001 > \ldots \ldots \ldots > 0$$

And we know 0 is not positive. So no matter how small positive number we may come up with, we can still make one the smaller. So there is no smallest positive number.

5.   What is the smallest negative number?

There is no smallest negative number.   Why not?

A number is negative if it is less than 0.

And the larger the magnitude, the smaller the number if the number is negative.

And the magnitude of the negative number can be infinitely large.

So a negative number can be infinitely small.

$$0 > -0.000001 > -0.001 > -0.1 > -1 > -5 > -99 > -9999 > -99999999 > \ldots \ldots \ldots$$

And thus, no matter how small negative number we may come up with, we can still make one the smaller. Therefore, there is no smallest negative number.

6.   What is the largest number where there are ten digits and two of the digits are below (to the right of) the point?

The number is 99,999,999.99.   How come?

To begin with, the number we want is positive. And the larger the digits, the larger the number if the number is positive.

And we know each digit is one of the integers from 0 to 9. So the largest digit can be 9. And next, in the number, 8 digits are above the point, and 2 digits are below the point.

So the number we want is 99,999,999.99.

7.   What is the largest positive number where there are ten digits, and seven of the digits are above (to the left of) the point?

The number is 9,999,999.999.   How come?

First, the number we want is positive. And the larger the digits, the larger the number if the number is positive. And we know each digit is one of the integers from 0 to 9. So the largest digit can be 9.

And next, the number has 7 digits above the point, and 3 digits below the point.
So the number we want is 9,999,999.999.

8.   What is the largest negative number where there are ten digits, and four of the digits are below the point?

The number is −100,000.0001.   How come?

First, the number we want is negative.
And the smaller the magnitude, the larger the negative number.

So the smaller the digit, the smaller the magnitude, so the larger the negative number.

And we know that the smallest number we can use as a digit in a number is 0.

And we know that the number we want is made of 10 digits, of which seven are above the point, and three are below the point. To what then, can we set the digits in the number we want?

We cannot set all the digits to 0s, because setting all to 0s, we just get 0.

The second smallest number we can use as a digit is 1.

So we want to set the highest digit to 1, and put the negative sign in front.

Then, we get: -1000000.0000, which is however, the same as -1000000, which does not have four digits below the point. So it is not the number we want.

What do we mean by a number that has 4 digits below the point?

At least, the lowest digit, that is, the fourth digit below the point is not 0.
So for instance, a number with 4 digits below the point can be 0.0128 or 12.0105.

What then, is the number we want?

In the number, the highest digit is 1, the lowest digit is 1, too, and the rest of all the digits are 0s. So the number we want is $-100,000.0001$.

And let's now put all the ideas together in one page.

8. What is the largest negative number where there are ten digits, and four of the digits are below the point?

First, the number we want is negative.

And the smaller the magnitude, the larger the negative number.

So the smaller the digit, the smaller the magnitude, so the larger the negative number.

And we know that the smallest number we can use as a digit is 0, and that the number is made of 10 digits, of which seven are above the point, and three are below the point. To what then, can we set the digits in the number we want?

We cannot set all the digits to 0s, because setting all to 0s, we just get 0.

And the second smallest number we can use as a digit is 1.

So we want to set the highest digit to 1, and put the negative sign in front.

Then, we get: −1000000.0000, which is however, the same as −1000000, which does not have four digits below the point. So it is not the number we want.

What do we mean by a number that has 4 digits below the point?

At least, the lowest digit, that is, the fourth digit below the point is not 0.

So for instance, a number with 4 digits below the point can be 0.0128 or 12.0105.

What then, is the number we want?

In the number, the highest digit is 1, the lowest digit is 1, too, and the rest of all the digits are 0s. So the number we want is −100,000.0001.

9. What is the smallest number where there are twelve digits, and three of those are below the point?

The number is −999,999,999.999.   How come?

First, the number we want is negative. And the larger the digits, the smaller the number if the number is negative. And we know each digit is an integer from 0 to 9. So the largest digit can be 9.

And next, the number has 9 digits above the point, and 3 digits below the point.
So the number we want is –999,999,999.999.

A.   What is the smallest positive number that has five digits above the point and seven digits below the point?

The number is 10,000.0000001.   How come?

First, the number we want is positive.
And the smaller the magnitude, the smaller the positive number.
So the smaller the digit, the smaller the magnitude, so the smaller the positive number.

And we know that the smallest number we can use as a digit is 0, and that the number we want is made of 12 digits, of which 5 are above the point, and 7 are below the point.
To what then, can we set the digits in the number we want?

We cannot set all the digits to 0s, because setting all to 0s, we just get 0.
And the second smallest number we can use a digit is 1.
So we want to set the highest digit to 1.

Then, we get: 10000.0000000, which is however, no other than 10000, which does not have 7 digits below the point. So it is not the number we want.
What do we mean by a number that has 7 digits below the point?

At least, the lowest digit, that is, the seventh digit below the point is not 0.
So for instance, a number with 7 digits below the point can be 101.0117008.
What then, is the number we want?

In the number, the highest digit is 1, the lowest digit is 1, too, and the rest of all the digits are 0s.   So the number we want is 10,000.0000001.

And let's now put all the ideas together in one page.

A.   What is the smallest positive number that has five digits above the point and seven digits below the point?

First, the number we want is positive.

And the smaller the magnitude, the smaller the positive number.

So the smaller the digit, the smaller the magnitude, so the smaller the positive number.

And we know that the smallest number we can use as a digit is 0, and that the number we want is made of 12 digits, of which 5 are above the point, and 7 are below the point.

To what then, can we set the digits in the number we want?

We cannot set all the digits to 0s, because setting all to 0s, we just get 0.

And the second smallest number we can use a digit is 1.

So we want to set the highest digit to 1.

Then, we get: 10000.0000000, which is however, no other than 10000, which does not have 7 digits below the point. So it is not the number we want.

What do we mean by a number that has 7 digits below the point?

At least, the lowest digit, that is, the seventh digit below the point is not 0.

So for instance, a number with 7 digits below the point can be 101.0117008.

What then, is the number we want?

In the number, the highest digit is 1, the lowest digit is 1, too, and the rest of all the digits are 0s. So the number we want is 10,000.0000001.

B.   What is the smallest negative number that has 10 digits above the point and two digits below the point?

The number is –9,999,999,999.99.   How come?

First, the larger the digits, the smaller the number if the number is negative. And we know each digit is an integer from 0 to 9. So the largest digit can be 9.

And next, the number has 10 digits above the point, and 3 digits below the point.

So the number we want is –9,999,999,999.99.

C. What is the largest 10-digit number where all digits are different, and three of the digits are above the point?

   The number is 987.6543201.   How come?

First, the number we want is positive.
And the larger the magnitude, the larger the positive number.
So the larger the digit, the larger the magnitude, so the larger the positive number.

We know however, all the digits in the number we want are different.
To what then, can we set the digits in the number we want?

First, assuming A is the number we want, and ignoring the point, that is, the dot in A, we can set: $A = 9876543210$.
And next, we know that three digits are above the point.
So can we just set: $A = 987.6543210$?

We know: $A = 987.6543210 = 987.654321$, which does not have 10 digits.
So we want to set: $A = 987.6543201$.

132

D. **What is the largest 10-digit negative number where all digits are different, and five of the digits are below the point?**

The number is −10234.56789.   How come?

The smaller the magnitude, the larger the negative number.
So the smaller the digit, the smaller the magnitude, so the larger the negative number.

We know however, all the digits in the number we want are different.
To what then, can we set the digits in the number we want?

First, assuming A is the number we want, and ignoring the point, that is, the dot in A, we can try setting: $A = -0123456789$, which is however, no other than **−123456789**.
Can we then, set: $A = -1234567890$?

The smaller the digit, the smaller the magnitude, so the larger the negative number.
So we want to set: $A = -1023456789$.

And next, we want A to be the smallest of all the negative numbers that have 5 digits above the point and 5 digits below the point.

So the number we want is: **−10234.56789**.

E. **What is the smallest 12-digit number where five digits are above the point, and every digit above the point is different from any of the digits below the point?**

The number is −99999.8888888.   How come?

First, the number we want is negative, since a number negative is smaller than a number positive. And the larger the magnitude, the smaller the negative number.

So the larger the digit, the larger the magnitude, so the smaller the negative number. Thus first, assuming A is the smallest 12-digit number with no digit below the dot, we can set: $A = -999999999999$.

Next, we know in the number we want, 5 digits are above the point and every digit above the point is different from any of the digits below the point.
And the second largest number we can use as a digit is 8.
So the number we want is: -99999.8888888.

F. **Find the smallest 12-digit positive number where 7 digits are above the point, and every digit above the point is different from any of the digits below the point.**

The number is 1,000,000.22222.   How come?

The smaller the magnitude, the smaller the positive number.
The smaller the digit, the smaller the magnitude, so the smaller the positive number.

Thus first, assuming A is the smallest 12-digit positive number with no digit below the dot, we can set: $A = 100000000000$.

Next, we know in the number we want, 7 digits are above the point and every digit above the point is different from any of the digits below the point.

So we cannot use 1 as any digit below the dot.
And the second smallest number we can use as a digit is 2.

So the number we want is 1,000,000.22222.

G.   Find the smallest 12-digit number where six digits are above the point, and are different, and every digit above the point is different from any digit below the point.

The number is –987654.333333.   How come?

First, the number we want is negative.

And the larger the magnitude, the smaller the negative number.

So the larger the digit, the larger the magnitude, so the smaller the negative number.

Thus next, assuming *A* is the smallest 12-digit number with no digit below the dot, to what number can we set *A*?

We can set: *A* = **–999999999999**.

Next, we know in the number we want, 6 digits are above the point, and are different.

So assuming *B* is the smallest 12-digit number where 6 digits are above the point, and are different, to what number can we set *B*?

We can set: *B* = **–987654.999999**.

And next, we know in the number we want, every digit above the point is different from any of the digits below the point.

So we cannot use any digit above the dot as any digit below the dot.

What then, are the numbers can we use as digits below the dot?

They are 0, 1, 2, and 3.

And we know: the larger the magnitude, the smaller the negative number.

So the larger the digit, the larger the magnitude, so the smaller the negative number.

And 3 is the largest of 3, 2, 1, and 0.

So the number we want is –987654.333333.

Sense of Arithmetic 1

How many dots in each color do you find in each rectangle below?

And how many dots does each rectangle have?

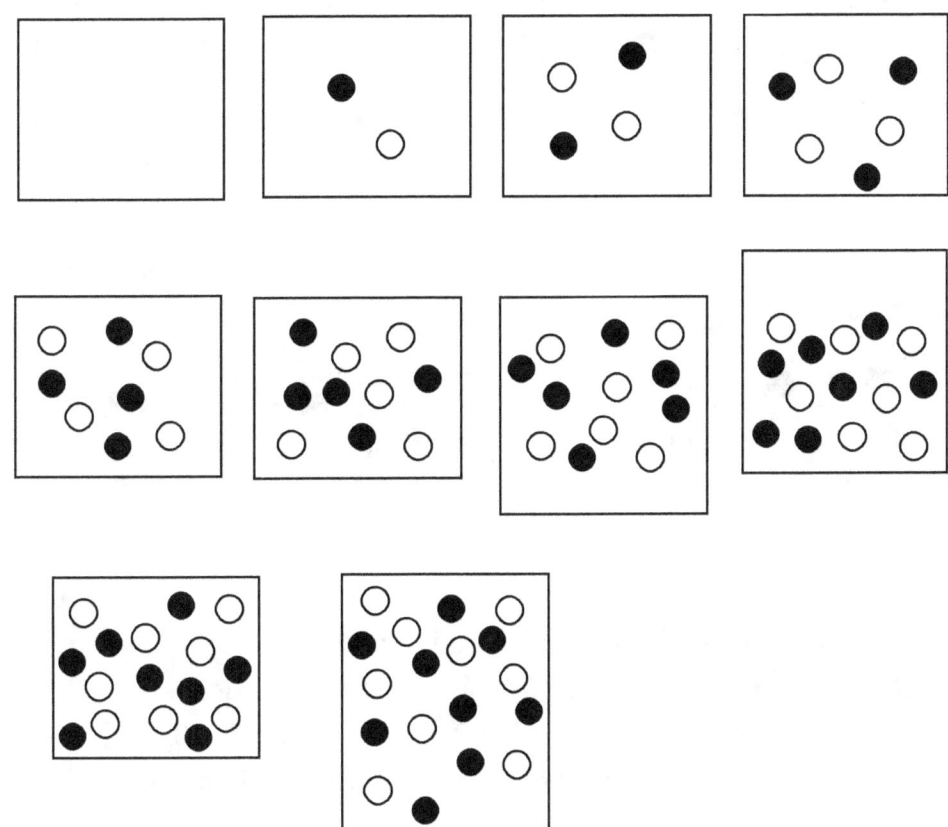

How many dots in each color do you find in each rectangle below?

And how many dots does each rectangle have?

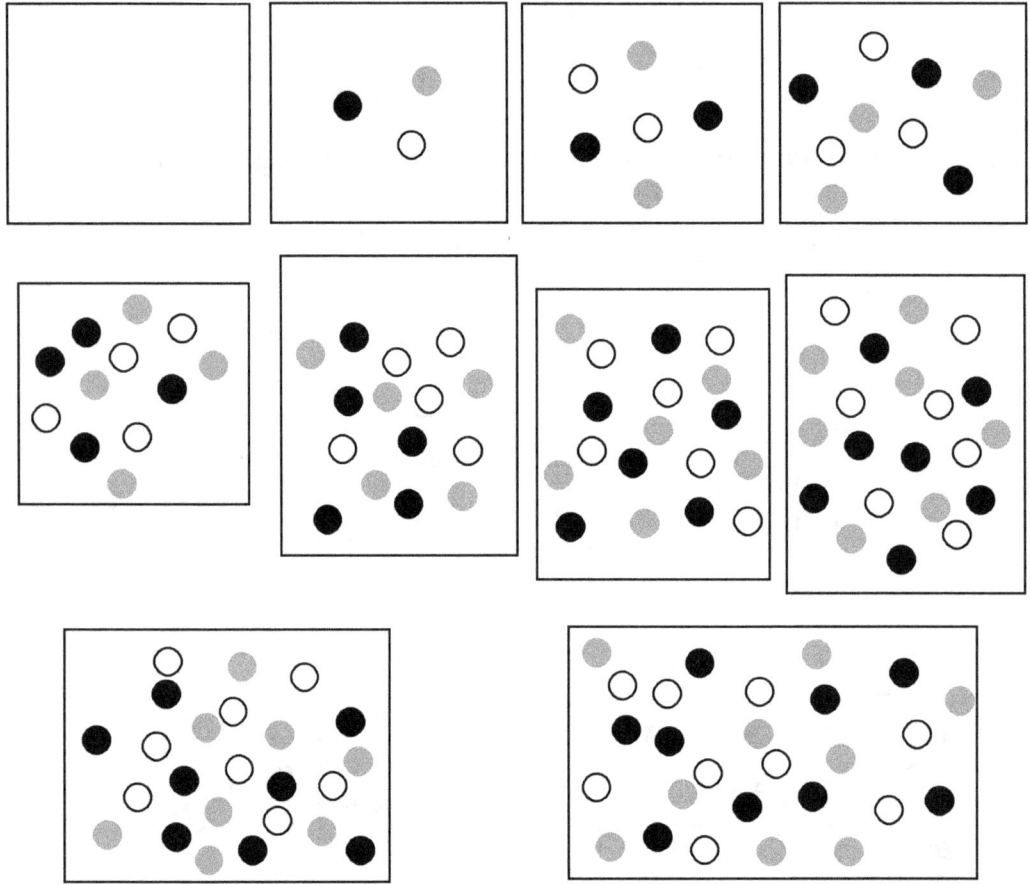

Read the calculations below if want to. And if reading them, just keep
reading at your pace. It can help increase sense of arithmetic.
You don't have to read all at once. Each reading can take 5, 10, or 15
minutes at a time, or as much as you can concentrate. And having read them
all, read them again when you feel like it.
If however, getting used to much of the stuff, move on to the next section.

Note however, you don't have to do calculations the way below.
Each calculation just shows that **there are many ways we can calculate**.

$0 = 1 \times 0 = 0 \times 1 = 0 + 0 = 2 \times 0 = 0 \times 2 = 0 + 0 + 0 = 3 \times 0 = 0 \times 3$

$= 0 + 0 + 0 + 0 + \ldots + 0 = 0 \times 97 = 97 \times 0 = 238 \times 0 = 0 = \ldots$

$1 = 1 \times 1 = 1 + 1 - 1 = 2 - 1 = 3 + 2 - 2 - 3 + 1 = 0 + 1 = 1 + 0 = \ldots$

$2 = 1 + 1 = 1 \times 2 = 2 \times 1 = \ldots$

3

$= 1 + 1 + 1 = 1 \times 3 = 3 \times 1$

$= 1 + 2 = 1 + (1 + 1) = 1 + 1 \times 2 = 1 + 2 \times 1$

$= 2 + 1 = (1 + 1) + 1 = 1 \times 2 + 1 = 2 \times 1 + 1$

$= 0 + 3 = \ldots$

4

$= 1 + 1 + 1 + 1 = 1 \times 4$

$= 1 + 3 = 1 + 1 \times 3$

$= 3 + 1 = 1 \times 3 + 1$

$= 2 + 2 = 2 \times 2 = \ldots$

5

$= 1 + 1 + 1 + 1 + 1 = 1 \times 5 = 5 \times 1$

$= 1 + 4 = 4 + 1 = 2 + 3 = 3 + 2 = \ldots$

6

= 1 x 6 = 6 x 1

= 1 + 5 = 5 + 1

= 2 + 4 = 2 + 2 x 2

= 4 + 2 = 2 x 2 + 2

= 2 + 2 + 2

= 2 x 3 = 3 x 2 = . . .

7

= 1 x 7 = 7 x 1

= 2 + 5 = 5 + 2

= 1 + 6 = 1 + 2 x 3 = 1 + 3 x 2

= 3 + 4 = 3 + 2 x 2 = 1 x 3 + 1 x 2 x 2

= 1 x 1 x 3 + 1 x 2 x 2 x 1 = . . .

8

= 8 x 1

= 1 + 1 + 1 + . . . + 1 = 1 x 8

= 2 + 2 + 2 + 2 = 2 x 4

= 4 + 4 = 4 x 2

= 3 + 5 = 5 + 3 = . . .

9

$= 9 \times 1$

$= 1 + 1 + 1 + \ldots + 1 = 1 \times 9$

$= 4 + 5 = 5 + 4$

$= 4 + (4 + 1) = 4 + 4 + 1 = 4 \times 2 + 1 = 2 \times 4 + 1$

$= 8 + 1 = 2 \times 4 + 1 = 4 \times 2 + 1$

$= 3 + 3 + 3 = 3 \times 3$

$= 3 + 6 = 3 + (3 + 3) = 3 + 3 \times 2 = 3 \times 1 + 3 \times 2 = 3 \times (1 + 2) = 3 \times 3$

$= 6 + 3 = (3 + 3) + 3 = 3 \times 2 + 3 = 3 \times 2 + 3 \times 1 = 3 \times (2 + 1) = 3 \times 3$

$= 2 + 7 = 7 + 2$

$= 2 + (2 + 5)$

$= 2 + 2 + 5$

$= 2 \times 2 + 5$

$= 2 \times 2 + (2 + 3)$

$= 2 \times 2 + 2 + 3$

$= 2 \times 2 + 2 \times 1 + 3$

$= 2 \times (2 + 1) + 3$

$= 2 \times 3 + 3$

$= 3 \times 2 + 3 \times 1$

$= 3 \times (2 + 1)$

$= 3 \times 3$

$= (2 + 1) \times (2 + 1)$

$= (2 + 1) \times 2 + (2 + 1) \times 1$

$= (2 \times 2 + 1 \times 2) + (2 \times 1 + 1 \times 1)$

$= 2 \times 2 + 2 + 2 + 1$

$= 4 + 2 + 2 + 1$

$= 4 + 2 \times 2 + 1$

$= 4 + 4 + 1 = 4 \times 2 + 1 = \ldots$

$10 = 10 \times 1 + 1 \times 0$

$= 1 + 9 = 9 + 1$
$= 2 + 8 = 8 + 2 = 8 + 1 + 1$
$= 3 + 7 = 7 + 3 = 7 + 2 + 1$
$= 4 + 6 = 6 + 4 = 6 + 2 + 2 = 6 + 2 \times 2$
$= 5 + 5 = 5 \times 2 = 2 \times 5$
$= 1 + 9 = 1 \times 1 + 9 \times 1 = 1 \times 1 + 1 \times 9$

$= 1 + 1 + 8 = 1 \times 2 + 8 \times 1 = 2 \times 1 + 1 \times 8$
$= 1 + 1 + 1 + 7 = 1 \times 3 + 7 \times 1 = 3 \times 1 + 1 \times 7 = 1 + 2 + 7$

$= 1 + 1 + 1 + 1 + 6 = 2 + 2 + 6$
$= 2 \times 2 + 3 + 3 = 3 \times 2 + 2 \times 2 = 2 \times 3 + 2 \times 2 = 2 \times (3 + 2) = 2 \times 5$

$= 1 + 1 + 1 + 1 + 1 + 5 = 1 \times 5 + 5 \times 1 = 5 \times 1 + 5 \times 1 = 5 \times (1 + 1)$
$= 5 \times 2 = 2 \times 5$

$= (1 + 1) + (1 + 1) + (1 + 1) + (1 + 1) + (1 + 1) = 2 + 2 + 2 + 2 + 2$
$= 2 \times 5 = 5 \times 2$

$= (1 + 1 + 1 + 1 + 1) + (1 + 1 + 1 + 1 + 1) = 5 + 5 = 5 \times 2 = 2 \times 5$

$= 1 + 1 + 1 + 1 + 1 + 1 + 1 + 1 + 1 + 1$
$= 1 + (1 + 1) + (1 + 1 + 1) + (1 + 1 + 1 + 1)$
$= 1 + 2 + 3 + 4 = (1 + 2) + (3 + 4) = 3 + 7$

$= 1 + 2 + 3 + 4 = 2 + 4 + 1 + 3 = 2 + (4 + 1) + 3 = 2 + 5 + 3$

$= 2 + (5 + 3) = 2 + 8 = \ldots$

$11 = 10 + 1 = 5 \times 2 + 1 = 5 + 5 + 1 = 1 \times 5 + 1 \times 5 + 1 = 1 \times (5 + 5 + 1)$
$= 1 \times 11 = \ldots$

12
$= 10 + 2$
$= 5 \times 2 + 2$
$= 2 \times 5 + 2 \times 1$
$= 2 \times (5 + 1)$
$= 2 \times 6$
$= 6 \times 2$
$= 6 + 6$
$= 4 + 2 + 4 + 2 = 4 \times 2 + 2 + 2 = 4 \times 2 + 4 = 4 \times (2 + 1) = 4 \times 3 = 3 \times 4 = \ldots$

$13 = 10 + 3 = 5 \times 2 + 3 = 5 \times 2 + 2 + 1 = 12 + 1 = \ldots$

$14 = 10 + 4 = 5 \times 2 + 4 = 2 \times 5 + 2 \times 2 = 2 \times (5 + 2) = 2 \times 7 = 7 \times 2 = \ldots$

$15 = 10 + 5 = 5 \times 2 + 5 = 5 \times (2 + 1) = 5 \times 3 = 5 \times (1 + 1 + 1) = 5 + 5 + 5$
$= \ldots$

16
$= 10 + 6$
$= 10 + 5 + 1$
$= 5 \times 2 + 5 + 1$
$= 5 + 5 + 5 + 1$

$= 5 + 1 + 4 + 5 + 1$

$= 6 + 4 + 6$

$= 6 \times 2 + 4$

$= 12 + 4$

$= (10 + 2) + 4$

$= 10 + 2 + 4$

$= 5 + 5 + 2 + 4$

$= 4 + 1 + 4 + 1 + 2 + 4 = 4 + 4 + 4 + 1 + 1 + 2 = 4 + 4 + 4 + 4 = 4 \times 4$

$= 4 \times 3 + 1 + 1 + 2$

$= 4 \times 3 + 4$

$= 4 \times (3 + 1)$

$= 4 \times (1 + 2 + 1)$

$= 4 \times 1 + 4 \times 2 + 4 \times 1$

$= 4 + 8 + 4$

$= 4 + 12 = \ldots$

$17 = 10 + 7 = 5 \times 2 + 7 = 5 \times 2 + (5 + 2) = 5 \times 3 + 2 = \ldots$

18

$= 10 + 8 = 5 \times 2 + 8 = 5 \times 2 + 4 \times 2 = (5 + 4) \times 2 = 9 \times 2 = 2 \times 9 = \ldots$

$= 6 + 4 + 8$

$= 6 + 4 + 6 + 2$

$= 6 + 6 + 4 + 2$

$= 6 + 6 + 6 = 6 \times 3 = 3 \times 6 = \ldots$

$19 = 10 + 9 = 5 \times 2 + 5 + 4 = 5 \times 3 + 4 = \ldots$

$20 = 10 \times 2 = (5 \times 2) \times 2 = 5 \times 2 \times 2 = 5 \times 4 = 4 \times 5 = \ldots$

$21 = 10 \times 2 + 1 = 10 + 10 + 1 = 10 + 11 = \ldots$

$22 = 20 + 2 = 10 \times 2 + 1 \times 2 = (10 + 1) \times 2 = 11 \times 2 = 2 \times 11 = \ldots$

$23 = 20 + 3 = 10 \times 2 + 1 \times 3 = \ldots$

$\ldots$

$98 = 10 \times 9 + 8 = 90 + 8 = 90 + 4 + 4 = 90 + 4 \times 2 = \ldots$

$99 = 10 \times 9 + 9 = (5 + 5) \times 9 + 9 = (5 + 5 + 1) \times 9 = 11 \times 9 = \ldots$

$100 = 100 \times 1 + 10 \times 0 + 1 \times 0$

$100 = 100 \times 1 = 10 \times 10 \times 1 = 10 \times 10$

$100 = 0 + 100 = 1 + 99 = 2 + 98 = 3 + 97 = \ldots$

$100 = 0 + 100 = 5 + 95 = 10 + 90 = 15 + 85 = 20 + 80 = \ldots$

$$
\begin{aligned}
100 &= 10 + 20 + 30 + 40 \\
&= 10 \times 1 + 10 \times 2 + 10 \times 3 + 10 \times 4 \\
&= 10 \times (1 + 2 + 3 + 4) \\
&= 10 \times 10
\end{aligned}
$$

$$
\begin{aligned}
&= 5 \times 2 + 5 \times 4 + 5 \times 6 + 5 \times 8 \\
&= 5 \times (2 + 4 + 6 + 8) \\
&= 5 \times 20 = \ldots
\end{aligned}
$$

Note again, you don't have to do calculations the way above.

Each calculation just shows that **there are many ways we can calculate**.

And there can be many ways you can break a value into pieces, and put the pieces together to get the same value.   How many ways, though?

As many as you can think of.

Sense of Arithmetic 2

Read the calculations below if you want to. And if reading them, just keep reading at your pace. It can help increase sense of arithmetic. You don't have to read all at once. Each reading can take 5, 10, or 15 minutes at a time, or as much as you can concentrate. And having read them all, read them again when you feel like it.

<u>**Note however**</u>, you don't have to do calculations the way below.
Each calculation just shows that **there are many ways we can calculate**.

$0 + 0 = 0 \times 2 = 0$, and $0 - 0 = 0$

$0 + 1 = 1$, and $0 - 1 = -1$

$1 - 0 = 1$, $1 + 1 = 1 \times 2 = 2$, and $1 - 1 = 0$

$2 + 1 = 1 \times 2 + 1 = 1 + 1 + 1 = 1 \times 3 = 3$

$2 - 1 = 1 + 1 - 1 = 1 + 0 = 1$

$1 - 2 = 1 - (1 + 1) = 1 - 1 - 1 = 0 - 1 = -1$

$3 + 2 = 1 \times 3 + 1 \times 2 = 1 \times (3 + 2) = 1 \times 5 = 5$

$3 - 2 = 1 + 2 - 2 = 1 + 0 = 1$

$2 - 3 = 2 - (2 + 1) = 2 - 2 - 1 = 0 - 1 = -1$

$5 + 3 = 8$, $5 - 3 = 2$, and $3 - 5 = -2$

$8 + 4$

$= 8 + (2 + 2) = (8 + 2) + 2 = 10 + 2 = 12$

$= (2 + 6) + 4 = 2 + (6 + 4) = 2 + 10 = 12$

$8 - 4 = 4$

$10 + 0 = 10 \times (1 + 0) + 1 \times (0 + 0) = 10 \times 1 = 10$

$10 - 0$

$= 10 \times (1 - 0) = 10$

$= 10 \times (1 - 0) + 1 \times (0 - 0) = 10 + 0 = 10$

$10 + 1 = 11$

$10 - 1 = (0 + 1 \times 10) - 1 = 0 + (1 \times 10 - 1) = 0 + 9 = 9$

$1 - 10$

$= 1 - (0 + 1 \times 10) = 1 - (0 + 1 + 1 \times 9) = 1 - 1 - (0 + 1 \times 9)$

$= 0 - (0 + 9) = 0 - 9 = -9$

$10 + 9 = 19$

$10 - 9 = (0 + 1 \times 10) - 9 = (0 + (1 + 9)) - 9 = 1 + 9 - 9 = 1 + 0 = 1$

$10 + 10 = 10 \times 2 = 20$

$10 - 10 = 10 \times (1 - 1) = 10 \times 0 = 0$

$10 + 11 = 21$

$10 - 11 = 10 - (10 + 1) = 10 - 10 - 1 = 0 - 1 = -1$

$11 - 10 = (10 + 1) - (10 + 0) = 10 \text{ x } (1 - 1) + 1 \text{ x } (1 - 0) = 0 + 1 = 1$

$10 + 19 = 29$

$10 - 19 = 10 - (10 + 9) = (10 - 10) + (0 - 9) = 0 + (-9) = -9$

$19 - 10 = (10 + 9) - 10 = (10 - 10) + (9 - 0) = 0 + 9 = 9$

$14 + 4 = 18$

$14 - 4 = 10 \text{ x } (1 - 0) + 1 \text{ x } (4 - 4) = 10 \text{ x } 1 + 1 \text{ x } 0 = 10 + 0 = 10$

$4 - 14 = 10 \text{ x } (0 - 1) + 1 \text{ x } (4 - 4) = 10 \text{ x } (-1) + 1 \text{ x } 0 = 10 \text{ x } (-1) + 0 = -10$

$16 + 9$
$= 16 + (4 + 5)$
$= (16 + 4) + 5$
$= \{(10 + 6) + 4\} + 5$
$= \{10 + (6 + 4)\} + 5$
$= (10 + 10) + 5$
$= 10 \text{ x } 2 + 5 = 20 + 5$

$16 + 9$
$= (10 + 6) + 9$
$= \{10 + (5 + 1)\} + 9$
$= \{(10 + 5) + 1\} + 9$
$= (10 + 5) + (1 + 9)$
$= (10 + 5) + 10$
$= 10 + 5 + 10 = 10 + 10 + 5 = 10 \text{ x } 2 + 5$

16 – 9

= 10 x (1 – 0) + 1 x (6 – 9) = 10 x 1 + (-3) = 10 + (-3) = 10 – 3 = 7

= 10 x (0 – 0) + 1 x (10 + 6 – 9)

= 10 x 0 + 1 x (6 + 10 – 9) = 0 + 1 x (6 + 1) = 1 x 7 = 7

9 – 16

= 10 x (0 – 1) + 1 x (9 – 6)

= 10 x (-1) + 1 x 3

= -10 + 3

= 3 + (-10)

= 3 – 10 = 1 x (3 – 10) = 1 x (-7) = -7

-10 + 3

= -(7 + 3) + 3

= 3 + {-(7 + 3)}

= 3 + {-(3 + 7)}

= 3 – (3 + 7)

= 3 – 3 – 7 = 0 – 7 = -7

39 + 58

= (30 + 9) + (50 + 8)

= 30 + 50 + 9 + 8

= 30 + 50 + 7 + (2 + 8)

= 30 + 50 + 7 + 10

= 90 + 7 = 97

$30 + 50 + 7 + 10$

$= 10 \times (3 + 5 + 1) + 7 = 10 \times (8 + 1) + 7 = 10 \times 9 + 7 = 90 + 7 = 97$

$39 - 58$

$= 10 \times (3 - 5) + 1 \times (9 - 8)$

$= 10 \times (-2) + 1 \times 1$

$= 10 \times (-1) + ((-1) \times 10 + 1)$

$= -10 + (1 + (-1) \times 10)$

$= -10 + (1 + (-1) \times (1 + 9))$

$= -10 + (1 + ((-1) + (-9)))$

$= -10 + (1 + (-1)) + (-9)$

$= -10 + (1 - 1) + (-9)$

$= -10 + 0 + (-9)$

$= -10 + (-9)$

$= (-10) \times 1 + (-1) \times 9 = -19$

$78 + 85 + 69$

$= (70 + 80 + 60) + (8 + 5 + 9)$

$= (70 + 80 + 60) + (8 + (2 + 3) + 9)$

$= (70 + 80 + 60) + ((10 + 3) + 9)$

$= (70 + 80 + 60) + (10 + (3 + 9))$

$= (70 + 80 + 60) + (10 + ((2 + 1) + 9))$

$= (70 + 80 + 60) + (10 + (2 + 10))$

$= (70 + 80 + 60) + (10 \times 2 + 2)$

$= 10 \times (7 + 8 + 6 + 2) + 1 \times 2$

$= 10 \times (15 + 6 + 2) + 1 \times 2$

$= 10 \times (21 + 2) + 1 \times 2$

$= 10 \times 23 + 1 \times 2$

$= 10 \times (10 \times 2 + 3) + 1 \times 2$

$= 100 \times 2 + 10 \times 3 + 1 \times 2$

$= 200 + 30 + 2 = 232$

$98 - 24 - 58 = (98 - 24) - 58 = 74 - 58 = 16$

$(98 - 24) - 58 = 98 - (24 + 58) = 98 - 82 = 16$

$34 - 45 - 76 + 27$

$= 34 - (45 + 76) + 27$

$= 34 + 27 - (45 + 76)$

$= 61 - 121$

$= (60 + 1) + ((-100) + (-20) + (-1))$

$= -100 + (60 + (-20)) + (1 + (-1))$

$= -100 + 40 + 0$

$= -100 \times 1 + 10 \times 4 + 1 \times 0$

$= -100 \times 0 + (10 \times 4 + (-10) \times 10) + 0$

$= 0 + 10 \times 4 + 10 \times (-10) + 0 = 10 \times (4 + (-10)) = 10 \times (-6) = -6 \times 10 = -60$

$= 34 + 27 - 45 - 76 = 61 + (-(45 + 76)) = 61 + (-121)$

$= 61 + (-61) + (-60) = -60$

0 x 0 = 0

0 x 1 = 0   0 x (-1) = 0

1 x 0 = 0   (-1) x 0 = 0

3 x 2 = 3 + 3 = 6   (-3) x 2 = -3 + (-3) = -6

3 x (-2) = 3 x (-1) x 2 = -1 x 3 x 2 = -1 x 6 = -6

-3 x (-2) = -1 x 3 x (-1) x 2 = -1 x (-1) x 3 x 2 = 1 x 6 = 6

12 x 3 = 12 + 12 + 12 = 36

= (10 + 2) x 3 = 10 x 3 + 2 x 3 = 30 + 6 = 36

15 x 12
= (10 + 5) x 12
= 10 x 12 + 5 x 12
= 10 x (10 + 2) + 5 x (10 + 2)
= 10 x 10 + 10 x 2 + 5 x 10 + 5 x 2
= 100 + 20 + 50 + 10 = 180

0.1 + 0.1 = 0.2   0.1 − 0.1 = 0

4.4 + 32.7
= (1 x 4 + 0.1 x 4) + (10 x 3 + 1 x 2 + 0.1 x 7)
= 10 x (0 + 3) + 1 x (4 + 2) + 0.1 x (4 + 7)
= 10 x 3 + 1 x (4 + 2) + 0.1 x (10 + 1)
= 10 x 3 + 1 x (4 + 2 + 1) + 0.1 x 1
= 30 + 7 + 0.1 = 37.1

4.4 – 32.7

= (1 x 4 + 0.1 x 4) – (10 x 3 + 1 x 2 + 0.1 x 7)

= 10 x (0 + (-3)) + 1 x (4 + (-2)) + 0.1 x (4 + (-7))

= 10 x (-3) + 1 x 2 + 0.1 x (-3)

= 10 x ((-2) + (-1)) + 1 x 2 + 0.1 x (-3)

= 10 x (-2) + 10 x (-1) + 1 x 2 + 0.1 x (-3)

= 10 x (-2) + (-10) + 1 x 2 + 0.1 x (-3)

= 10 x (-2) + 1 x (-10) + 1 x 2 + 0.1 x (-3)

= 10 x (-2) + 1 x ((-10) + 2) + 0.1 x (-3)

= 10 x (-2) + 1 x (-8) + 0.1 x (-3) = -28.3

1.8 x 2 = 1.8 + 1.8 = 1 x (1 + 1) + 0.1 x (8 + 8)

= 1 x (1 + 1) + 0.1 x (10 + 6)

= 1 x (1 + 1 + 1) + 0.1 x 6 = 1 x 3 + 0.1 x 6 = 3.6

1.8 + 1.8

= (1 + 0.8) x 2

= 1 x 2 + 0.8 x 2

= 1 x 2 + 0.1 x 8 x 2

= 1 x 2 + 0.1 x 16

= 1 x 2 + 0.1 x (10 + 6)

= 1 x2 + 0.1 x 10 + 0.1 x 6

= 1 x (2 + 1) + 0.1 x 6

= 1 x 3 + 0.1 x 6 = 3.6

$0.3 \times 0.2 = 0.1 \times 3 \times 0.1 \times 2 = 0.1 \times 0.1 \times 3 \times 2 = 0.01 \times 6 = 0.06$

0.3×0.8

$= 0.1 \times 3 \times 0.1 \times 8$

$= 0.1 \times 0.1 \times 3 \times 8$

$= 0.01 \times 24$

$= 0.01 \times (20 + 4)$

$= 0.01 \times 20 + 0.01 \times 4$

$= 0.01 \times 2 \times 10 + 0.01 \times 4$

$= 0.01 \times 10 \times 2 + 0.01 \times 4$

$= 0.1 \times 2 + 0.01 \times 4$

$= 0.2 + 0.04 = 0.24$

1.9×0.7

$= (1 + 0.9) \times 0.7$

$= 1 \times 0.7 + 0.9 \times 0.7$

$= 0.7 + 0.1 \times 0.1 \times 9 \times 7$

$= 0.7 + 0.01 \times (60 + 3)$

$= 0.7 + 0.01 \times (10 \times 6 + 3)$

$= 0.7 + (0.01 \times 10) \times 6 + 0.03$

$= 0.7 + 0.1 \times 6 + 0.03$

$= 0.7 + 0.6 + 0.03$

$= 0.1 \times (7 + 6) + 0.03$

$= 0.1 \times (10 + 3) + 0.03$

$= 0.1 \times 10 + 0.3 + 0.03$

$= 1 + 0.3 + 0.03 = 1.33$

5.9 x 9.8

$= (5 + 0.9) \text{ x } (9 + 0.8)$

$= 5 \text{ x } (9 + 0.8) + 0.9 \text{ x } (9 + 0.8)$

$= 5 \text{ x } 9 + 5 \text{ x } 0.8 + 0.9 \text{ x } 9 + 0.9 \text{ x } 0.8$

$= 45 + 5 \text{ x } (0.1 \text{ x } 8) + 0.1 \text{ x } (9 \text{ x } 9) + 0.1 \text{ x } 0.1 \text{ x } (9 \text{ x } 8)$

$= 45 + 5 \text{ x } (8 \text{ x } 0.1) + 0.1 \text{ x } 81 + 0.01 \text{ x } 72$

$= 45 + (5 \text{ x } 8) \text{ x } 0.1 + 0.1 \text{ x } 81 + 0.01 \text{ x } 72$

$= 45 + 40 \text{ x } 0.1 + 0.1 \text{ x } 81 + 0.01 \text{ x } 72$

$= 45 + 0.1 \text{ x } 40 + 0.1 \text{ x } 81 + 0.01 \text{ x } 72$

$= 45 + 0.1 \text{ x } (40 + 81) + 0.01 \text{ x } 72$

$= 45 + 0.1 \text{ x } 121 + 0.01 \text{ x } 72$

$= 40 + 5 + 0.1 \text{ x } (100 + 20 + 1) + 0.01 \text{ x } (70 + 2)$

$= 40 + 5 + 0.1 \text{ x } 100 + 0.1 \text{ x } 20 + 0.1 + 0.01 \text{ x } 70 + 0.01 \text{ x } 2$

$= 40 + 5 + 0.1 \text{ x } 10 \text{ x } 10 + 0.1 \text{ x } 10 \text{ x } 2 + 0.1 + 0.01 \text{ x } 10 \text{ x } 7 + 0.01 \text{ x } 2$

$= 40 + 5 + 10 + 2 + 0.1 + 0.1 \text{ x } 7 + 0.01 \text{ x } 2$

$= 50 + 7 + 0.1 \text{ x } (1 + 7) + 0.01 \text{ x } 2$

$= 50 + 7 + 0.1 \text{ x } 8 + 0.01 \text{ x } 2$

$= 57.82$

0.04 x 7.9

= 0.04 x (7 + 0.9)

= 0.04 x 7 + 0.04 x 0.9

= (0.01 x 4) x 7 + (0.01 x 4) x 0.9

= 0.01 x (4 x 7) + (0.01 x 4) x (0.1 x 9)

= 0.01 x (4 x 7) + 0.01 x 0.1 x 4 x 9

= 0.01 x 28 + 0.001 x 36

= 0.01 x 28 + 0.001 x (30 + 6)

= 0.01 x 28 + 0.001 x (10 x 3 + 6)

= 0.01 x 28 + 0.01 x 3 + 0.001 x 6

= 0.01 x (28 + 3) + 0.001 x 6

= 0.01 x 31 + 0.006

= 0.01 x (30 + 1) + 0.006

= 0.01 x (10 x 3) + 0.01 x 1 + 0.006

= 0.1 x 3 + 0.01 x 1 + 0.006

= 0.3 + 0.01 + 0.006

= 0.316

0 / 1 = 0 / 2 = 0 / 3 = 0 / 10 = 0 / 26 = 0 / 502 = 0.

And in fact, we have: 0 / A = 0 for all A ≠ 0.

1 / 1 = 1 since 1 = 1 x 1

-1 / 1 = -1 since -1 = 1 x (-1) = (-1) x 1

1 / (-1) = -1 since 1 = (-1) x (-1) = 1

-1 / (-1) = 1 since -1 = (-1) x 1 = -1

2 / 1 = 2 since 2 = 1 x 2

-2 / 1 = -2 since -2 = 1 x (-2) = (-2) x 1

2 / (-1) = -2 since 2 = (-1) x (-2) = (-1) x ((-1) x 2) = ((-1) x (-1)) x 2 = 1 x 2

2 / 2 = 1 since 2 = 2 x 1

-2 / 2 = -1 since -2 = 2 x (-1) = (-1) x 2 = -2

2 / -2 = -1 since 2 = (-2) x (-1) = (-1) x 2 x (-1) = (-1) x (-1) x 2 = 1 x 2

4 / 2 = 2 since 4 = 2 x 2, and 6 / 3 = 2 since 6 = 3 x 2

10 / 1 = 10 since 10 = 1 x 10, and 12 / 3 = 4 since 12 = 3 x 4

By the way, no division by 0 is allowed.
So no denominator in any fraction is 0. Why not?

First, 0 / 0 = 0, 1, 7, or any number, because if 0 / 0 = A, we get: 0 = 0 x A for all A. So no consistency, and thus, no solution.

Next, assuming A / 0 = B for some A ≠ 0 and B ≠ 0, we get: A = 0 x B, which is 0 no matter what B may be, but we have assumed that A ≠ 0. So we can use no number as B, and thus, A / 0 is not possible.

28 / 2
= (20 + 8) / 2 = (20 / 2) + (8 / 2) = 10 + 4 = 14
= (30 − 2) / 2 = (30 / 2) − (2 / 2) = 15 − 1 = 14

1248 / 4

= (1200 + 40 + 8) / 4

= (1200 / 4) + (40 / 4) + (8 / 4)

= ((100 x 12) / 4) + ((10 x 4) / 4) + ((1 x 8) / 4)

= (100 x (12 / 4)) + (10 x (4 / 4)) + (1 x (8 / 4))

= 100 x 3 + 10 x 1 + 1 x 2

= 300 + 10 + 2 = 312

1248 / 4 / 2

= (1248 / 4) / 2

= 1248 / (4 x 2)

= 1248/8

= 156

1248 / 4 / 2

= 312 / 2

= (100 x 3 + 10 x 1 + 1 x 2) / 2

= 100 x (3 / 2) + 10 x (1 / 2) + 1 x (2 / 2)

= 100 x (1 + 1 / 2) + 10 x (1 / 2) + 1 x (2 / 2)

= 100 x 1 + 100 / 2 + 10 / 2 + 1 x 1

= 100 x 1 + (10 x 10) / 2 + (1 x 10) / 2 + 1 x 1

= 100 x 1 + 10 x (10 / 2) + 1 x (10 / 2) + 1 x 1

= 100 x 1 + 10 x 5 + 1 x 5 + 1 x 1

= 100 x 1 + 10 x 5 + 1 x (5 + 1)

= 100 x 1 + 10 x 5 + 1 x 6

= 156

1 / 2

= (0.1 x 10) / 2 = 0.1 x (10 / 2) = 0.1 x 5 = 0.5

= 0.5 since 0.5 x 2 = 1

0.1 / 2

= (0.01 x 10) / 2 = 0.01 x (10 / 2) = 0.01 x 5 = 0.05

= 0.05 since 0.05 x 2 = 0.1

3 / 2 = (2 + 1) / 2 = 2 / 2 + 1 / 2 = 1 + 0.5 = 1.5

15 / 6

= (12 + 3) / 6

= 12 / 6 + 3 / 6

= 2 + (1 x 3) / 6

= 2 + ((0.1 x 10) x 3) / 6

= 2 + (0.1 x (10 x 3)) / 6

= 2 + (0.1 x 30) / 6

= 2 + 0.1 x (30 / 6)

= 2 + 0.1 x 5

= 2 + 0.5 = 2.5

3.9 / 2

= (3 + 0.9) / 2

= (1 x 3 + 0.1 x 9) / 2

= (1 x 3) / 2 + (0.1 x 9) / 2

= 1 x (3 / 2) + 0.1 x (9 / 2)

= 1 x ((2 + 1) / 2) + 0.1 x ((8 + 1) / 2)

= 1 x ((2 / 2) + (1 / 2)) + 0.1 x ((8 / 2) + (1 / 2))

= 1 x (1 + 0.5) + 0.1 x (4 + 0.5)

= 1 x 1 + 0.5 + 0.4 + 0.05

= 1 x 1 + 0.1 x (5 + 4) + 0.01 x 5

= 1 x 1 + 0.1 x 9 + 0.01 x 5

= 1.95

2 / 4

= (1 x 2) / 4

= ((0.1 x 10) x 2) / 4

= (0.1 x (10 x 2)) / 4

= (0.1 x 20) / 4

= 0.1 x (20 / 4)

= 0.1 x 5

= 0.5

1 / 4

= (0.1 x 10) / 4 = 0.1 x (10 / 4) = 0.1 x ((8 + 2) / 4) = 0.1 x ((8 / 4) + (2 / 4))

= 0.1 x (2 + 0.5) = 0.1 x 2 + 0.05 = 0.1 x 2 + 0.01 x 5 = 0.25

9.7 / 4

= (9 + 0.7) / 4

= 9 / 4 + 0.7 / 4

= (8 + 1) / 4 + (0.1 x 7) / 4

= 2 + 1 / 4 + 0.1 x (7 / 4)

= 2 + 0.25 + 0.1 x ((4 + 3) / 4)

= 2 + 0.25 + 0.1 x (1 + 3 / 4)

= 2 + 0.25 + 0.1 x (1 + (0.1 x 30) / 4)

= 2 + 0.25 + 0.1 x (1 + 0.1 x (30 / 4))

= 2 + 0.25 + 0.1 x (1 + 0.1 x ((28 + 2) / 4))

= 2 + 0.25 + 0.1 x (1 + 0.1 x (7 + 2 / 4))

= 2 + 0.25 + 0.1 x (1 + 0.1 x (7 + 0.5))

= 2 + 0.25 + 0.1 x (1 + 0.7 + 0.05)

= 2 + 0.25 + 0.1 + 0.07 + 0.005

= 2 + 0.2 + 0.05 + 0.1 + 0.07 + 0.005

= 2 + 0.1 x (2 + 1) + 0.01 x (5 + 7) + 0.005

= 2 + 0.1 x 3 + 0.01 x (10 + 2) + 0.005

= 2 + 0.1 x 3 + 0.01 x 10 + 0.01 x 2 + 0.005

= 2 + 0.1 x 3 + 0.1 x 1 + 0.01 x 2 + 0.005

= 2 + 0.1 x 4 + 0.01 x 2 + 0.005

= 2.425

$1 / 0.1 = (0.1 \times 10) / 0.1 = (10 \times 0.1) / 0.1 = 10 \times (0.1 / 0.1) = 10 \times 1 = 10$

$1 / 0.2$

$= (0.1 \times 10) / 0.2$

$= (10 \times 0.1) / 0.2$

$= 10 \times (0.1 / 0.2)$

$= 10 \times ((0.01 \times 10) / 0.2)$

$= 10 \times (0.01 \times (10 / 0.2))$

$= 10 \times (0.01 \times ((0.1 \times 100) / 0.2))$

$= 10 \times (0.01 \times ((0.1 \times (2 \times 50)) / 0.2))$

$= 10 \times (0.01 \times (((0.1 \times 2) \times 50) / 0.2))$

$= 10 \times (0.01 \times ((0.2 \times 50) / 0.2))$

$= 10 \times (0.01 \times ((50 \times 0.2) / 0.2))$

$= 10 \times (0.01 \times (50 \times (0.2 / 0.2)))$

$= 10 \times (0.01 \times (50 \times 1))$

$= 10 \times (0.01 \times (10 \times 5))$

$= 10 \times (0.1 \times 5)$

$= (10 \times 0.1) \times 5$

$= 1 \times 5$

$2 / 0.4$

$= (0.1 \times 20) / 0.4 = (0.1 \times (4 \times 5)) / 0.4 = ((0.1 \times 4) \times 5) / 0.4$

$= (0.4 \times 5) / 0.4 = (5 \times 0.4) / 0.4 = 5 \times (0.4 / 0.4) = 5 \times 1$

7.9 / 0.4

= (7 + 0.9) / 0.4

= 7 / 0.4 + 0.9 / 0.4

= (4 + 3) / 0.4 + (0.1 x 9) / 0.4

= 4 / 0.4 + 3 / 0.4 + 0.1 x (9 / 0.4)

= 10 + (1 x 3) / 0.4 + 0.1 x (1 x 9) / 0.4

= 10 + ((0.1 x 10) x 3) / 0.4 + 0.1 x 0.1 x 10 x 9 / 0.4

= 10 + (0.1 x 30) / 0.4 + 0.1 x (0.1 x 90) / 0.4

= 10 + 0.1 x ((28 + 2) / 0.4) + 0.1 x 0.1 x (88 + 2) / 0.4

= 10 + 0.1 x (28 / 0.4 + 5) + 0.1 x 0.1 x (88 / 0.4 + 5)

= 10 + 0.1 x ((7 x 4) / 0.4 + 5) + 0.1 x 0.1 x (22 x 4 / 0.4 + 5)

= 10 + 0.1 x (7 x (4 / 0.4) + 5) + 0.1 x 0.1 x (22 x (4 / 0.4) + 5)

= 10 + 0.1 x (7 x 10 + 5) + 0.1 x 0.1 x (22 x 10 + 5)

= 10 + 0.1 x 7 x 10 + 0.1 x 5 + 0.1 x (0.1 x (22 x 10) + 0.1 x 5)

= 10 + 7 + 0.5 + 0.1 x (22 + 0.5)

= 10 + 7 + 0.5 + 0.1 x (20 + 2)+ 0.05

= 10 + 7 + 0.5 + 0.1 x (2 x 10) + 0.2 + 0.05

= 10 + 7 + 0.5 + 2 + 0.2 + 0.05 = 10 + 9 + 0.7 + 0.05 = 19.75

Note again, you don't have to do calculations the way above.
Each calculation just shows that **there are many ways we can calculate**.

And there can be many ways you can break a value into pieces, and put the pieces together to get the same value as many ways as you can think of.

www.ingramcontent.com/pod-product-compliance
Lightning Source LLC
Chambersburg PA
CBHW081448170526
45166CB00008B/2355